D1121460

The Macroscope

The Macroscope

A NEW WORLD SCIENTIFIC SYSTEM

by Joël de Rosnay

Translated from the French by Robert Edwards

Harper & Row, Publishers

New York, Hagerstown, San Francisco, London

 For information address Harper & Row, Publishers, Inc., 10 East 53rd Street, New York, N.Y. 10022. Published simultaneously in Canada by Fitzhenry & Whiteside Limited, Toronto.

FIRST EDITION

Designed by Janice Stern

Library of Congress Cataloging in Publication Data

de Rosnay, Joël
The macroscope.

Translation of Le macroscope.
Bibliography: p.
Includes index.
1. Social systems. 2. System theory. 3. Energy policy. 4. Communication. 5. Time. I. Title.
H61.R6813 301.1 76-5122
ISBN 0-06-011029-5

79 80 81 82 83 84 10 9 8 7 6 5 4 3 2 1

To Stella, Tatiana, Cécilia and Alexis

Contents

Acknowledgments

I should like to thank here all those who have given me their advice and their help in the conception and the editing of this book.

First, for their criticism and suggestions at the time of the reading of the manuscript or of certain chapters, Jean-Jacques Balan, Madeleine Barthélémy-Madaule, Georges Guéron, Pierre Jablon, Jacques Monod, Olivier de Nervaux, Massimo Piatelli-Palmarini, François de Rougemont, Jean-Claude Roumanteau.

The fruitful discussions of the Group of Ten, led by Jacques Robin, have permitted me to elaborate on my train of thought and to compare it to that of others. I want especially to thank for their advice, support, and constructive criticism, Henri Atlan, Jacques Attali, Henri Laborit, Edgar Morin, and René Passet.

Thanks to the understanding of Jean Guéroult and Georges Guéron, I have been able to find conditions that allowed me to carry on simultaneously the writing of the book and the pursuit of my professional duties.

Stella has been not only my advisor at every moment and the laboratory of my ideas but also a valued assistant.

Finally, I owe particular thanks to Catherine Fourneau-Faye for the successive typings of the manuscript, her advice, and her encouragement; to Anne Boissel-Puybaraud for her understanding in the execution of the drawings; to Babette Roumanteau for her careful readings and rereadings of the manuscript and her judicious corrections of its style.

Introduction: The Macroscope

Microscope, telescope: these words evoke the great scientific penetrations of the infinitely small and the infinitely great. The microscope has permitted a dizzying plunge into the depths of living matter; it has made possible the discovery of the cell, microbes, and viruses; it has advanced the progress of biology and medicine. The telescope has opened the mind to the immensity of the cosmos; it has traced the path of the planets and the stars and has prepared men for the conquest of space.

Today we are confronted with another infinite: the infinitely complex. We are confounded by the number and variety of elements, of relationships, of interactions and combinations on which the functions of large systems depend. We are only the cells, or the cogs; we are put off by the interdependence and the dynamism of the systems, which transform them at the very moment we study them. We must be able to understand them better in order to guide them better. And this time we have no instrument to use. We have only our brain—our intelligence and our reason—to attack the immense complexity of life and society. True, the computer is an indispensable instrument, yet it is only a catalyst, nothing more than a much-needed tool.

We need, then, a new instrument. The microscope and the telescope have been valuable in gathering the scientific knowledge of the universe. Now a new tool is needed by all those who would try to understand and direct effectively their action in this world, whether they are responsible for major decisions in politics, in science, and in industry or are ordinary people as we are.

I shall call this instrument the *macroscope* (from *macro,* great, and *skopein,* to observe).

The macroscope is unlike other tools. It is a symbolic instrument made of a number of methods and techniques borrowed from very different disciplines. It would be useless to search for it in laboratories and research centers, yet countless people use it today in the most varied fields. The macroscope can be considered the symbol of a new way of seeing, understanding, and acting (Fig. 1).

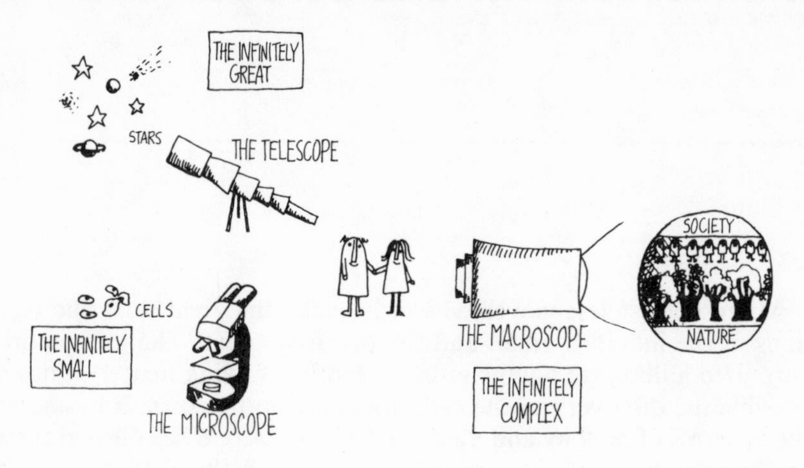

Fig. 1

Let us use the macroscope to direct a new look at nature, society, and man and to try to identify new rules of education and action. In its field of vision organizations, events, and evolutions are illuminated by a totally different light. The macroscope filters details and amplifies that which links things together. It is not used to make things larger or smaller but to observe what is at once too great, too slow, and too complex for our eyes (human society, for example, is a gigantic organism that is totally invisible to us). Formerly, in trying to comprehend a complex system, we sought the simplest units that explained matter and life: the molecule, the atom, elementary particles. Today, in relation to society, we are the particles. This time our glance must be directed toward the systems which surround us in order to better understand them before they destroy us. The roles are reversed: it is no longer the biologist who observes a living cell through a microscope; it is the cell itself that observes in the macroscope the organism that shelters it.

There is much talk today of the importance of a "vision of the whole" and of an "effort to synthesize." These attitudes are judged necessary to solve the complex problems of the modern world. Unfortunately, our education has not prepared us for this. Look at the list of university disciplines; they divide nature into so many private properties, each carefully isolated from the others. More simply, recall the basic education you received in school: English, mathematics, science, history and geography, government, or modern language—so many fragmented worlds, the vestiges of a scattered knowledge.

Should one hold on to the "analytic method," which isolates elements and variables in order to examine them one by one? While the experts are isolating, analyzing, and discussing, the technological upheavals and the cultural revolution impose new adaptations on society. The growing gap in time between the perception of global problems and the arrival at major decisions makes our analytic methods appear even more inept.

Another approach, a complementary one, exists and will often be discussed in this book. This is (I shall explain the choice of the term) the systemic approach. This new approach is symbolized by the macroscope. It emphasizes a global approach to the problems or to the systems that one studies, and it concentrates on the play of interactions among their elements. What practical bearing does it have on resolving global problems? Can it help us to enlarge our vision of the world, to better transmit knowledge, to free new values and new rules that can motivate and support action? This book tries to answer these questions. It is intended to be practical. Its organization, its pedagogy, its message rest on three principles: to increase one's ability to *see* better, to correlate facts in order to *understand* better, and to identify situations so that one may *act* better.

The general organization of *The Macroscope* is in the image of the approach that it advocates and describes: my medium is also my message.

This approach does not lend itself readily to conventional forms of communication. It has been necessary to "invent" even the organization of the book; to "invent" the means of communication that it intends to establish.

I have, therefore, turned my back on the classic organization of the "linear" book, in which ideas, developments, and chapters follow one another in sequential form. That is a corridorlike, tunnellike book, a one-way traffic in which one understands the end only when he has assimilated the facts given at the start.

I prefer the "intersecting" book to the linear book. You pick it up where you want; you pursue it according to your desires by following several simple and precise rules from the beginning. Thus, if you wish, you will be able to compose a book a la carte, one that corresponds as much as possible to your own interests and to what you hope to find in it. That is why the chapters and sections of *The Macroscope* are relatively independent modules, all of which play a part in leading toward the vision of the whole.

Fig. 2

For those who prefer to be guided, I will describe the "logical" approach. The book has the structure of a double cone. In the beginning one approaches the structure and mechanisms common to many systems of nature; one observes.

In this way one approaches the summit of the first cone; the general method allows one to connect the systems—the systemic approach. Then one moves into the second cone, to the applications and the diverse propositions or suggestions that I submit for consideration (Fig. 2).

The first chapter, *Through The Macroscope,* is essentially didactic but is presented in a way that I hope is original. Here I apply systemic principles in order to bring out the functioning of the principal systems of nature. It is also a primer for those who want to acquire rapidly the fundamentals of modern ecology, economics, and biology. From the ecosystem to our extraordinary universe, the stages of the voyage are: ecology and economics, the city and the organization, the organism and the cell.

The second chapter, *The Systemic Revolution: A New Culture,* is an introduction to and a training in the new method of approach to complexity. The purpose of this chapter is to unveil the basic concept of "systems," to bring to light fundamental laws, general principles, and the properties that relate the principal systems of nature. This is the key to the book; it tells how to use the macroscope.

In the three chapters that follow, *Energy and Survival, Information and the Interactive Society, Time and Evolution,* I shall try to apply the systemic approach to three fundamental fields of knowledge: energy, information, and time. These three chapters constitute the heart of the book because they illustrate its purpose. In fact energy, information, and time are the eternal elements on which all action depends—the sequence of all knowledge and meaning. To envisage their multiple implication at the physical, biological, social, and philosophical levels, I propose to use the macroscope.

The sixth chapter, *Values and Education,* considers how the global vision (particularly the new generation's views on nature, society, and man) can set forth new values, outline tomorrow's education, and bring out the features of a new type of emerging society.

These are the traits that the conclusion, in the form of a scenario, will try to set forth.

The last chapters necessarily return to the first, for the first chapter tries to apply the principles of a new form of education. In fact the basic elements of the systemic education treated in the sixth chapter (p. 192) are the very principles that I put into practice in the beginning of the book. You may, if you like, begin this book at the end, with the *Scenario for a World.* Or you can follow another route; if, for example, you are interested in biology and ecology, and if you know economics and business well enough, then read first the sections on the cell and the ecosystem. If you are familiar with cybernetics and the systemic approach, then go directly to the sections on energy, information, evolution, and time.

I should also like to "spread" a new form of communication throughout this book. *The Macroscope* is not a book that intends to popularize difficult scientific concepts, even though the first part, with its numerous illustrations, presents complicated subjects in a simple manner. The popularizing book concentrates on a given field of knowledge and tries to present it in a language accessible to everyone. To grasp a general concept that draws on several disciplines, to succeed in a personal synthesis of scientific, economic, and sociological facts, is often difficult. It is also difficult to unify a "mosaic" of views made popular by different approaches and different languages. The popular books allow the reader to be led by the author, the reader depending on the author to "take him by the hand" and help him over the difficult passages.

I am anxious for a new form of dialogue. Rather than supply portions of pure knowledge, I should like to stimulate inventive thought—the imagination. I prefer to make you use your capacities for reflection, intuition, and synthesis. To be sure, such a participation demands an effort; yet I am convinced that that is a form of communication that truly fulfills the mind.

The method used here is that of the enrichment of concepts. Consequently I give few definitions. The definition seems to me to be an easy solution, and I do not want to communicate by ready-made slogans, by conceptual "kits" ready to be assembled. To enrich and clarify a difficult concept one must return to the concept several times, now reviewing it in a different light, now placing it in a new context.

This particular form of dialogue necessarily implies—especially in the first part of the book—a new language of communication. The transmission of pure knowledge in neat little packages is often used in traditional analytical instruction. Yet one must also evoke and retranslate relationships between disciplines, movement, complexity, interdependence. One must depend on intuition, on creative thought, and (why not?) on subjective thought. Along with traditional discourse and verbal explanation, I believe in the virtues of diagrams, illustrations, models, analogies, and metaphors. Of course everyone knows that diagrams are "always false," that generalizations are "hasty," that models are "simplistic," that metaphors are "easy," and that analogies are "dangerous." In order to communicate idea and thinking in the most diverse fields, I have used the entire arsenal at my disposal; I leave it to you to make use of the guardrails of thought in the awkward spots.

I realize the danger of my enterprise. This book is ambitious because it touches on biology, ecology, economics, information systems, education, sociology, and even philosophy. If I have been naive enough to write it, it is because I believe that we express well only what we have lived.

A few years ago, in writing a little book on modern biology, I chose the global view of man who observes in "the interior of himself" the fantastic universe of the cell. It was then necessary to write the sequel, man looking "beyond himself" at the macrolife in which he is integrated and of which he is the element: business, the city, economics, the ecosystem. Man has lacked the tool with which to study the macrolife—the macroscope—and he will learn to use it only after a period of training. That training is what I will try to offer here.

This book may also appear to be superficial; in touching on all subjects, it may seem only to skim the principal subjects. It is not a study in depth of biology, economics, or ecology. It appears to pass over the problems posed by energy, communication, participation, acceleration, and evolution. It tries to reflect on knowledge and its limits. It outlines the main feature of a new education and an emerging society. Yet once again my approach is different. You can see many things by looking down on the continents from a satellite; is that a superficial view? I think not; first, because details invisible on land now appear, then, above all, because this vision of the world poses new questions and suggests further studies.

The specialist's book approaches in detail a small number of sectors cut arbitrarily from a greater whole. Talking about economy, he may dwell on inflation. Talking of the body, he may concentrate on the brain. Talking of business, he may put marketing before all else. This book, on the contrary, attempts to return all the principal elements to the system in which they belong and to consider them each in relation to the others.

This is no longer the method of the "generalist." I think that one should be careful of generalists; they often remain in the realm of ideas and do not attack the reality of facts. On the contrary, we need the help of specialists who have learned, through experience and exposure to other disciplines, to look beyond and to communicate. Should we call them "synthesizing" specialists? I don't know, but *The Macroscope* has been written with that perspective in mind.

Finally, I must say that I mistrust almost instinctively universal models that claim to contain everything and unitary theories that claim to explain everything. True, they correspond to that natural tendency of the human mind that wants to bring together, to rearrange, to unite. But it is just because they are so satisfying to read that they can be dangerous. A model of the world can lead to the worst intransigence: one sorts—and eliminates—everything that does not fit in this model. A unitary theory can lead to the worst smugness: why bother to hunt, to criticize, to invent? I reject every closed and sterile representation. The models I propose are only points of departure for reflection; in no case are they points of arrival. These models should be confronted with reality and especially with risk; they must be attacked, destroyed, and rebuilt. For they can only evolve in confrontation with and stimulus from the outside world—and this means action. It is through an incessant round trip between representation and action that a conceptual model will evolve. In this sense the macroscopic vision and the systemic approach, which are the web of the book, will be utilized in education and in action.

One

Through the Macroscope

Today the world is messages, codes, pieces of information. What dissection tomorrow will dislocate our objects in order to recompose them in a new space? What new Russian doll will emerge from it?—François Jacob

The atom, the molecule, the cell, the organism, and society fit one within the other like a series of Russian dolls. The largest of these dolls is the size of our planet; it contains the society of men and their economies, the cities and industries that transform the world, the living organisms and the cells which comprise them. One could continue in this way to open successive dolls as far as the elementary particles, but let us stop here.

The purpose of this preliminary exploration is twofold. First it is a matter of providing a "primer" in ecology, economics, and biology—disciplines that force us today to modify our ways of thinking. The three are not often united in a single approach—a situation that offers a risk but also an advantage. The risk is that you may find that the material dealing with the field you know best is too schematic, too simplistic. The advantage (which derives directly from the use of the macroscope) is that you will be able to discover, in other fields, new knowledge that may enrich and enlarge your own personal outlook.

Then it is a matter of introducing the concepts of "systems" and "systemic approach," the bases of the new culture of the concerned man of the twenty-first century. The opening of each doll exposes examples and practical aspects in advance of the general theory. (Remember, there is nothing to keep you from beginning with the second chapter, on systems, if you wish.)

1. ECOLOGY

All life on earth rests on the present or past functioning of the ecosystem, from the smallest bacteria to the deepest forests, from the fragile plankton of the oceans to man, his agriculture, and his industry. Thanks to the reserves of energy accumulated during the life of the world, the complex structures of society are maintained: large cities, industries, and communications networks.

The ecosystem is literally the house of life, and the science that studies it is ecology. This term was created in 1866 by the German biologist Ernst Haeckel from the Greek *oikos,* house, and *logos,* science. Ecology is concerned with the relations that exist between living beings and the milieu in which they live.

Yet the ecosystem is much more than merely the milieu in which one lives. In a way it is itself a living organism. Its giant cycles activate everything in the mineral world and the living world. Its biological power plants produce billions of tons of organic matter, matter that is stockpiled, distributed, consumed, recycled in the form of mineral elements, then reintroduced in the same factories, to be replenished with solar energy and to return through the cycles that maintain the life of every organization.

In what movements, what transformations does this "life" of the ecosystem manifest itself? It is shown in atmospheric circulation—winds, the movement of the clouds, precipitation, everything that could be seen by studying the earth at a distance. It is manifested in the flow of water—streams and rivers moving to the seas, the great ocean currents, the displacement of glaciers. It is seen in the movements of the earth's crust—earthquakes, volcanos, erosion, sedimentation, and, over a sufficiently long period, the formation of mountain chains. Finally, there are the life cycles in which the basic materials of living beings are perpetually made, changed, and circulated.

All these movements, displacements, and transformations require energy. Whatever their nature or their diversity of being, they draw this energy from three principal sources: solar radiation, energy from the earth's core (seismic or thermal), and gravity. Solar radiation is by far the most important source of energy, for it represents 99 percent of the energy balance of our planet. Even energy furnished by fossil fuels is nothing more than solar energy in storage.

Solar energy, then, powers the cycles of the ecosystem. To set a machine in motion in order to produce work, energy must run from a hot source to a cold "sink"—where it disappears forever. In the case of the sun-and-earth system the hot source is the flux of solar energy (radiation of short wavelengths), the cold sink the space between the stars. That space directly absorbs heat reflected from the earth as well as heat produced by geological, biological, and industrial processes that take place on earth. This reflection dissipates energy; it disperses it, disorganizes it, renders it unusable in the production of work. Thus between the sun, the earth, and the "black depths" of the terrestrial environment, there is an irreversible current, a waterfall of energy that flows from hot to cold.

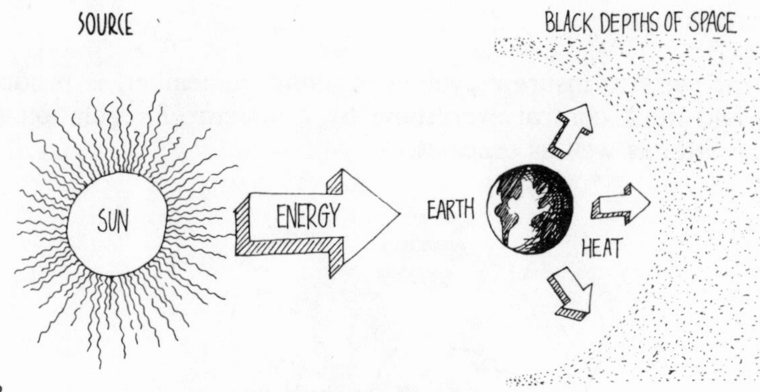

Fig. 3

In addition to the work produced on it (rapid movements and transformations), the earth holds its store of energy in equilibrium and consequently maintains its constant temperature through the radiation of heat toward space. It establishes a balance between energy received (energy used in the geological and biological processes) and energy broken down in irretrievable heat and irradiated toward space.* Only a negligible fraction of the immense quantity of solar energy received daily on the earth is used by living beings (Fig. 3).

The ecosystem is composed of four fields in strict interaction with one another: air, water, earth, and life. They are called respectively *atmosphere, hydrosphere, lithosphere,* and *biosphere.* The arrows in Figure 4 show that each field is related to all of the others. Even sediments on the floors of the oceans do not escape this rule; their composition depends not only on marine life and the composition of the oceans but also on the composition of the atmosphere.

The flow of energy that passes through the ecosystem is irreversible and inexhaustible. However, the chemical elements that make up all mineral or organic forms that we know on earth exist in finite number. These elements are found in the very heart of the ecosystem and are *recycled* after use. Everything that lives is made from building blocks that contain only six basic elements: carbon (C), hydrogen (H), oxygen (O), nitrogen (N), sulphur (S), and phosphate (P). The structures remain, but the elements of construction are replaced. Biologists call this dynamic renewal *turnover.* Living things (and the colonies they form—forests, populations, coral reefs) are continually being assembled and disassembled. Thus the ecosystem must have on hand a considerable stock of

---

* Energy broken down in irretrievable heat is called *entropy.* Later, I shall consider this important concept at greater length (see the third chapter).

replacement parts to insure recycling (nothing, remember, is produced *de novo*) and must control everything by a system of regulation that avoids scarcities as well as excesses.

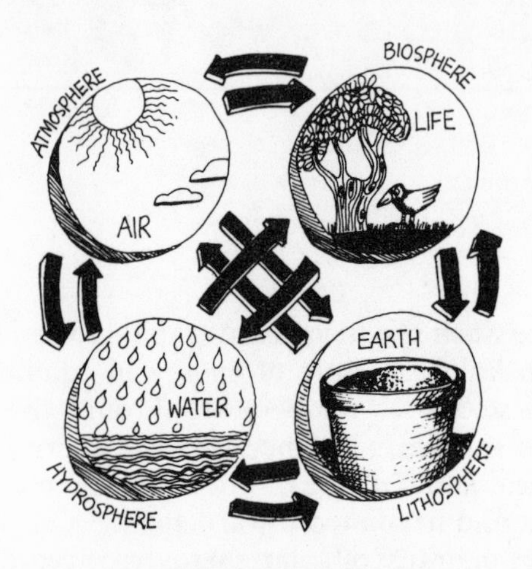

Fig. 4

To understand the finite nature of the ecosystem, one might imagine a bottle containing water, air, rocks, and a thin film of life (Fig. 5). When exposed to the sun, the bottle becomes the seat of great activity. The sun's rays, striking it at different angles and in different spots, cause inequalities of temperature and convection currents that produce movements of the air and the water. (The same thing happens on earth: similar differences of temperature set in motion formidable masses of air and water and produce winds, rain, waves, and currents.)

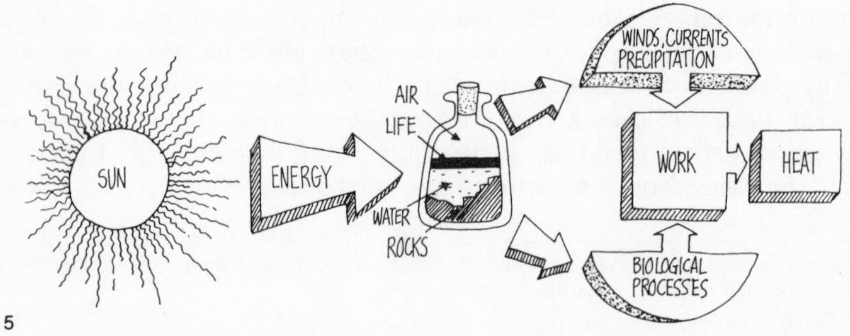

Fig. 5

If our bottle contains algae (simple one-cell plants capable of accomplishing photosynthesis) and protozoa (microscopic one-cell animals), the biological cycles can begin. The organic matter produced by the algae with the help of the sun's rays is "burned" by the protozoa. This combustion furnishes the energy that allows them, for example, to move about to look for food; it is made possible by the presence of oxygen released by the algae. And it is from carbon dioxide, the residue of the combustion, that the algae make organic material. The cycle is accomplished; all the elements in the bottle have been reused.

In the earth's ecosystem the elements essential to life are successively used and regenerated in the course of well-known cycles: the cycles of carbon, nitrogen, sulphur, and phosphate. These elements circulate among three huge reservoirs in which they are stockpiled: the reservoir of the atmosphere (and of the hydrosphere), the reservoir of the biomass (the mass of organic matter represented by all living beings), and the reservoir of sediments. Passing from one reservoir to another, the elements combine under different aspects: molecules of gas in the atmosphere, soluble ions (atoms that have lost or gained electrons) in the hydrosphere; crystalized salt in the sediments, organic molecules in the reservoirs of life.

In the atmosphere elements appear as molecules of gas: nitrogen (N_2), oxygen (O_2), sulphur dioxide (SO_2), and carbon dioxide (CO_2). In the hydrosphere, in sediments or in the soil, they occur in the form of soluble ions or bound in the form of salts (carbonates, nitrates, sulfates, phosphates).

The Economics of Nature: Production, Consumption, and Decomposition

The organic stage of the ecological cycles (the reservoir of life) can be considered the "motor" of all the cycles. In the course of this stage the principal substances responsible for the maintenance of life are made and consumed. Here the amount of oxygen in the atmosphere is regulated and the billions of tons of materials are stored. All of this comes about through an organization that is a model of industry and economy, with production, stockpiling, distribution, consumption, equitable sharing of energy, and complete recycling of materials. The three groups of organisms on which this industry and this economy rest are the *producers,* the *consumers,* and the *decomposers.*

The *producers* are green plants, aquatic vegetation, and, more generally, all organisms capable of *photosynthesis* (the production of organic material solely from solar light and mineral carbon gas). Producers are also called autotrophs.

The *consumers* are animals of all sizes, herbivores and carnivores, in terrestrial and aquatic milieus. They feed on living organisms and, through an internal oxidation called respiration, burn the organic materials that compose the tissues of their prey. Consumers are also called heterotrophs.

The *decomposers* feed on dead organisms or chemical substances dispersed in the environment.

Figure 6 sums up the relationship between these three groups whose activity makes possible the functioning of the ecosystem and the regulation of its equilibrium.

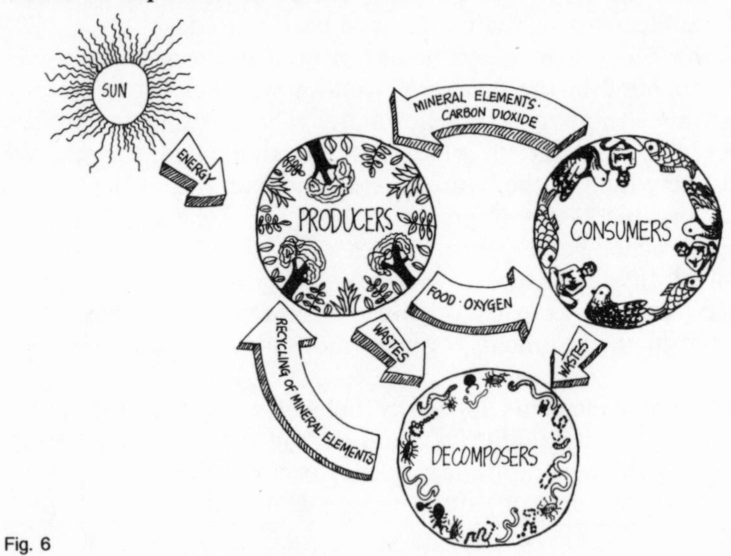

Fig. 6

During the day the producers manufacture great quantities of organic matter that accumulates in plant cells; at the same time there is an enormous output of oxygen. At night the process of oxidation takes over: the consumers oxidize (burn) the organic matter recently made and stored in order to produce the energy that allows them to perform work. This is the respiratory process. Of course animals, vegetation, and decomposers breathe during the day as well, but the animal processes of production are of such magnitude that they render virtually negligible the results of the oxidation that occurs simultaneously.

The two processes that form the basis of life, production and consumption (photosynthesis and respiration), are closely linked. An important difference exists between organisms capable of transforming radiant energy (light) and those that transform fixed energy, or energy trapped in the chemical combinations of organic molecules. Fixed energy is freed only when the combinations are broken. This happens during a free combustion (fire) or a controlled combustion (respiration).

Fixed energy then travels the length of the food chain (also called the *trophic chain*), which is made up of consumers (the herbivores and the various levels of carnivores). Each in turn profits to the maximum extent from the energy stored in the tissues of the organisms it captures—organisms that precede it in the series. This energy is used to the last particle at the time of the decomposition of animal bodies and vegetation. Microorganisms extract energy from already relatively simple organic molecules by transforming them into mineral molecules that are recirculated in the ecosystem.

Whatever the transforming organism may be, energy is lost to the food chain in three different ways.

Through *respiration,* since energy can produce work only by breaking down into irretrievable heat.
Through *consumption* by other organisms of vegetable or animal tissues that had stored energy.
Through *decomposition* of dead organisms and the elimination by vegetation and by animals of liquids and excrement.

Figures 7 and 8 give some idea of the flow of energy in the food chain and the losses incurred there.

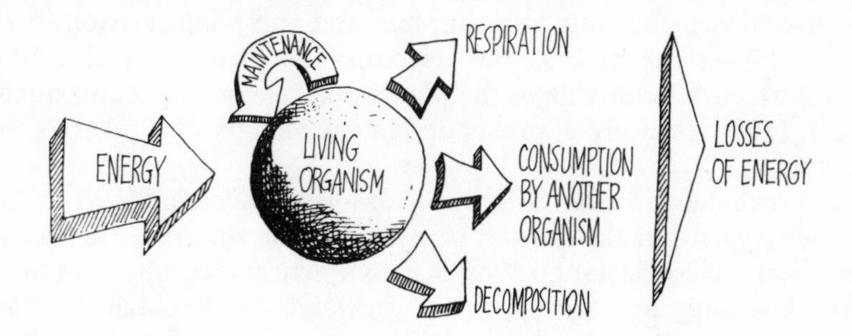

Fig. 7

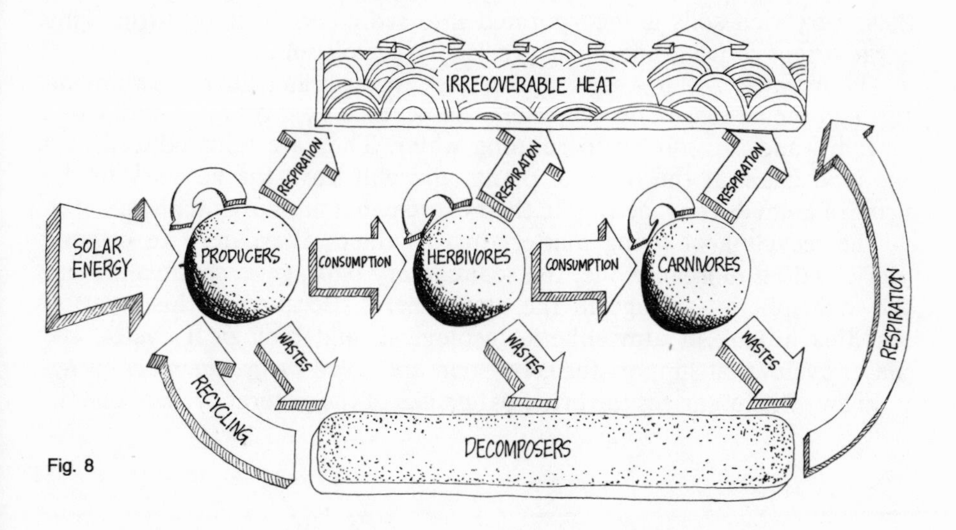

Fig. 8

If the role of the plant-producers and the animal-consumers is generally well known, that of the microorganic decomposers is much less so. However, it is due to the latter's prodigious activity that organic wastes are transformed into substances that are stored in sediments protected from oxidation. They then take the form of soluble molecules transported by running water or that of gaseous molecules liberated in the atmosphere. Thus all forms can be used again by the ecosystem over varying lengths of time.

What are the decomposers? They are bacteria, algae, fungi, yeast, protozoa, insects, mollusks, worms—a swarming population of minuscule beings with insatiable appetites. Organic molecules in excrement, urine, tissues in decomposition, and all degradable wastes are broken down by the decomposers into smaller and simpler fragments. This molecular breaking down leads, at the end of the series, to carbon dioxide and to water, the ultimate residue of the decomposition of organic material.

Complete decomposition is accomplished, for example, in the oxygen-rich environment of a forest or in a soil aerated by insects or turned over by earthworms. Residue that cannot be broken down forms humus. The mineral elements, nitrogen, sulphur, and phosphate, are totally regenerated. During the process the decomposers themselves also breathe; they return carbon dioxide to the plants and release important quantities of heat. (This is readily apprehended in the vicinity of a pile of compost or manure.)

The breaking down of molecules can also be accomplished in the absence of oxygen—at the bottom of a lake, in the slime of swamps, inside a dead body. Here decomposition is incomplete and combustion proceeds slowly, liberating less energy; this is *fermentation.* Residues that are incompletely burned accumulate on the spot and give off the singular odor of decaying matter (as in marshes, for example). The organic matter of these very rich soils is incorporated into sediments little by little. This is the origin of peat, coal, and eventually petroleum.

The nitrates, sulfates, and phosphates incorporated in the sediments through the actions of decomposers can be freed by erosion—wind, frost, or rain—and are soluble in running water. They are reintroduced into the food chain at the roots of plants and will leave the network in the urine of animals (nitrogen) or in their excrement (sulphur and phosphate).

The recycling of living matter thus produces alternately an organic phase and an inorganic one: the sedimentary (storage in sediments) and the atmospheric (storage in the atmosphere). Because of this rotation and this linking of atmospheric, geological, and biological cycles, the major cycles that support the ecosystem are called *biogeochemical cycles.*

Figure 9 summarizes the principal phases of the general cycle of chemical elements in the ecosystem (carbon, nitrogen, sulphur, and phosphate); it shows the circulation of chemical elements between the principal reservoirs. This diagram can be applied to each element, although certain cycles will have a predominant phase in the atmospheric reservoir or in the sedimentary reservoir.

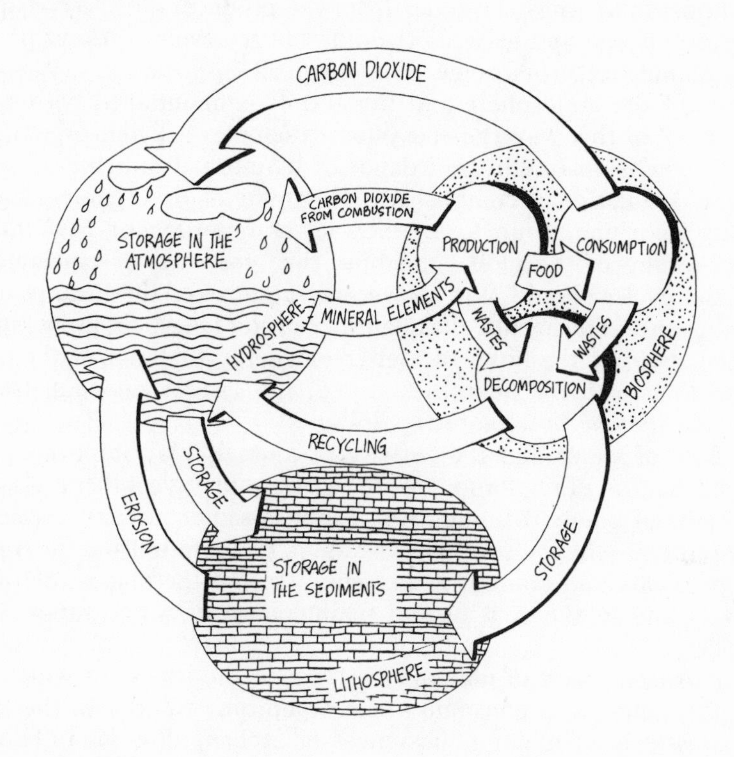

Fig. 9

Regulation and Maintenance of Equilibriums

The function of the ecosystem is not limited to the use of an irreversible flow of solar energy and to the cycles of production, storage, consumption, and regeneration of living matter. There is a third, equally important property: the regulation of the optimal functioning of the whole.

The biogeochemical cycles are self-regulating: a too extreme variation in one direction is immediately compensated by the modification of another variable, the overall effect of which is to return the system to its equilibrium. Each activity accomplished in the ecosystem has its counterpart. Each interaction, each exchange, no matter how minimal, is potentially a regulatory mechanism. The general effect of these mechanisms maintains the community in a "dynamic balance." Along each food chain or cycle the flows of energy and material travel. These chains and cycles are interconnected, coordinated, and synchronized in the greater assembly that constitutes the biogeochemical cycles.

The flow of materials continues from the producers to the consumers to the decomposers and between the different reservoirs without producing overabundance or shortage. The chemical composition of the great reservoirs of the atmosphere and the oceans is maintained within very strict limits. In this way the ecosystem resembles a living organism; it "knows" how to maintain the balance of its internal milieu.

How is regulation accomplished? The mineral or organic elements that pass from one group to another act as *activators* or *inhibitors* on the functioning of the global machine that produces or consumes. If one of the cycles should slow down, say because of the disappearance of certain consuming agents, the quantities in storage would grow rapidly. Since the speed of the flows of matter or energy that run in the cycles is proportionate to the quantities stored, the system balances itself by eliminating the overflow more rapidly.

The flow of water and the activity of animals play important roles in the mechanism of regulation. Water carries nutritive mineral elements to the roots of plants. Running water erodes sediments and accelerates their reentry into the cycles of the ecosystem. Evaporation and the transpiration of plants and animals are essential in the thermal regulation of organisms and in the control and maintenance of water vapor in the atmosphere.

The insatiable quest of animals for food, in the course of which they search for, catch, and consume other organisms, returns to the plants a regular stream of mineral substances in exchange for food. Thus the consumers work for the producers and producers work reciprocally for consumers. Each is "compensated" by the mineral elements or the food that the other group makes available to it. If the population of one kind of consumer increases too rapidly, the balance is upset, food becomes scarce, and individuals die of hunger—which reestablishes the optimum level of population for the setting in which this community lives.

The regulation of the size of a given population is based, then, on its struggle to obtain available food and on the mortality that strikes overabundant species through limited food sources. This regulatory mechanism is illustrated in Figure 10, the scheme of which I shall use often in the sections on economics and biology that follow.

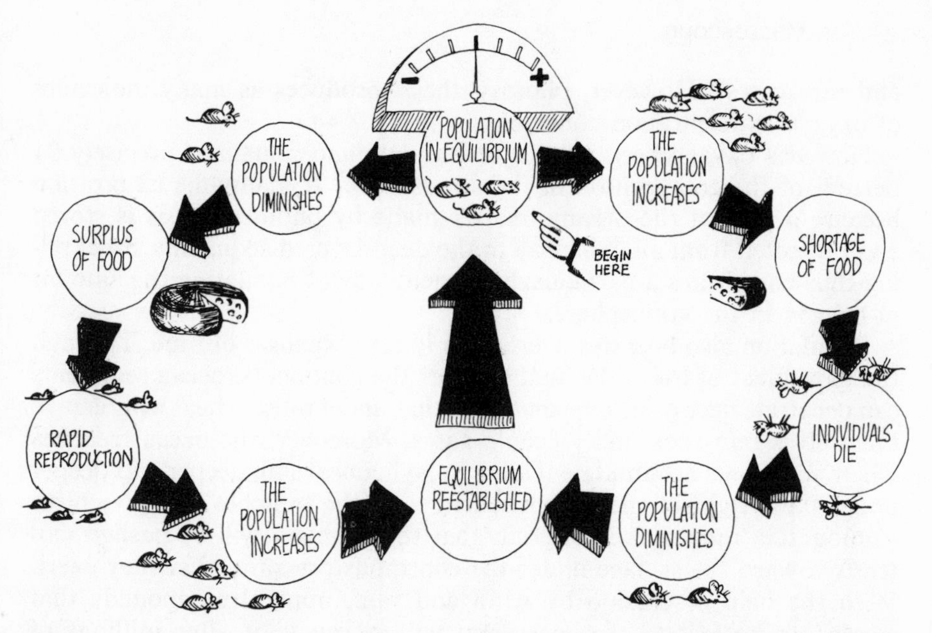

Fig. 10

Certain processes in the readjustment of equilibrium can be rapid, others extraordinarily slow. Ecologists have been able to measure, with the help of radioactive elements, the speed at which an element such as phosphate completes the organic cycle. They have seen how it passes from one organism to another, from chemical structure to chemical structure, from its entry into the food chain until its return to the mineral world. The complete period of the turnover (recycling) of phosphate, depending on the season, has actually been measured. In the case of a lake it ranges from ten minutes in summer to more than ten hours in winter; in the sedimentary stage the period of storage and liberation of phosphate can last two hundred years.

The three major reservoirs—the atmosphere, the hydrosphere, and the sediments—also play a regulatory role in the greater ecosystem by limiting the effects of sudden variations. They act as a buffer to reduce oscillations caused by cyclical variations. In this way the important concentration of carbonate ions in the ocean allows the almost constant concentration of carbon dioxide to be maintained in the atmosphere. Likewise, the interaction between atmosphere and sediments permits the regulation of the oxygen concentration in the atmosphere. This concentration has been maintained in a remarkable way at 21 percent for hundreds of millions of years (the rest of the atmosphere being 78 percent nitrogen and rare gases). However, photosynthesis produces as many molecules of oxygen as respiration consumes.

How has oxygen been able to accumulate and remain at precisely 21 percent of the composition of the atmosphere? It maintains its position because a part of the organic matter made by photosynthesis is stored and protected from all oxidation in the deep-layered sediments. Stockpiling thus constitutes a particularly efficient way of regulating the amount of oxygen in the atmosphere.

Regulation also operates over much longer expanses of time. Through the movement of the plates that support the continents, ocean sediments can penetrate deeply into the materials that under intense heat will remake themselves into rock and volcanic gases. Moreover, the ocean trenches where sediments accumulate (called geosynclines) sink deeper and deeper under the weight of the sediments. In time the trenches will give birth to mountain ranges; the materials that they contain will be pushed violently toward the surface under the enormous pressures that they exert. With the help of erosion by wind and rain, mineral compounds that seemed to be lost to the ecosystem will return to it after millions of years.

Thus the regulatory mechanisms between the mineral world and the biosphere, with their very different response times that may take from a minute to millions of years, allow the ecosystem to maintain its structure and its overall functions.

2. THE ECONOMY

In the ideal ecosystem that we have described so far, man is clearly missing. This new inhabitant of our planet has, by his agriculture, industry, and economy, modified little by little those equilibriums that existed long before him. Everything is happening as though a new organism (human society) were developing and growing from within the old. Man is like a parasite who takes energy and resources from his host and finally kills it.

How has man attained such power? He has no special instrument, but a number of means permit him to produce and distribute goods in constantly increasing quantities and on a larger and larger scale. It is the total of these means that constitutes the study of economics.

To study the general functioning of the economic machine through the macroscope one must adopt the outlook of a naturalist and observe from above the macroeconomic level. The economy is geared to the great ecological cycles—a fact that has long been forgotten or unrealized. When the economic machine accelerates or runs out of control, it consumes more energy, more material, more knowledge—and there will be more waste products dumped in the surrounding environment.

Such a point of view could lead to a naive interpretation of the economic machine if one did not keep in mind that behind the flows and cycles there are *centers of decision.* After all, it is in the midst of conflicts, oppositions, arbitrations, the search for power, and the domination of one group by another that one must restore the functioning of the economic machine. Yet such an approach would be beyond the scope of this book; our concern is not to describe a particular economic system (system in the political sense) but rather to outline, as we did for the ecosystem, the dynamics of the whole, the general functioning of the economic machine, whatever the system to which it may belong.

The word *economy* draws its meaning from the same roots as the word ecology. Economy (*oikos,* house, and *nomos,* rule) means literally the rule of conduct of a home. By extension it denotes the art of correctly managing one's goods and, in a limited sense, of managing one's goods by avoiding useless expenses or by effecting savings.

From the home, economic activity has extended to the state (political economy) and to society as a whole. The economic function of human society, in the broad sense, becomes the production of goods for the satisfaction of man's needs. The scarcity of goods and the difficulties of their production result in limitations on their distribution and use. We come, then, to the famous definition of L. Robbins: "Economics is the study of human behavior as a relation between rare means and ends which have mutually exclusive uses."

This kind of definition diminishes economic functions as well as the role of man (producer and consumer), who is motivated, it would seem, solely by the desire to satisfy his needs. Economics is then reduced, as François Perroux said, to a "science of means," the ends being the motives of morality and politics. One compares economics to the single function of a market where pure and perfect competition will reign. *Homo economicus* would appear to be a being without a soul, driven by rudimentary motives and barely capable of adapting passively to the "forces" of the market (René Passet).

This impoverishment of the economic function clearly appears in the classic diagrams of the economic cycles. They show a balance of forces between supply and demand, a flow of goods and services, a flow of money. Here is a machine that seems to be frozen, capable only of functioning by fits and starts, in an unreal universe from which nature is excluded (Fig. 11). The economic machine functions "between parentheses" without showing the irreversible flow of energy which inevitably breaks down in order to produce the work (see p. 2).

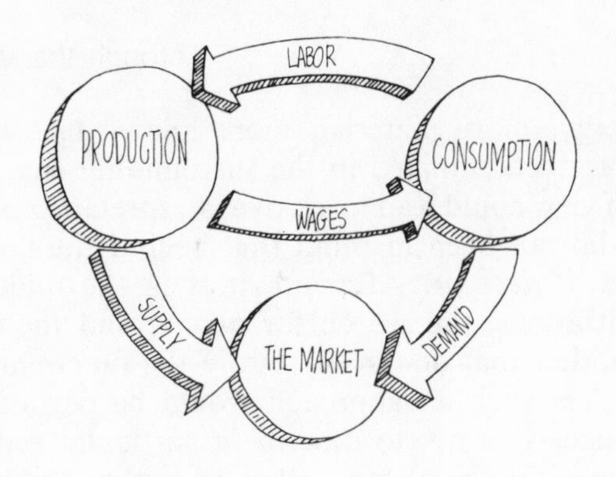

Fig. 11

Economy is also a "science of the living." To emphasize the close relationship between the ecosystem and the economic system, I should like to retrace, with the help of a series of diagrams, the brief history of the economy in its broadest sense. This is "the study of the mechanics of production, exchange, and consumption in a given social structure and the interdependencies between these mechanisms and this structure" (Attali and Guillaume).

A Short History of the Economy

The major stages in the development of the economic function coincide with the implementation of new powers that allow man to act more and more efficiently within his environment. Fire, agriculture, crafts and the perfection of tools, the steam engine, and the use of fossil fuels represent essential steps in the progress of man's domination of nature. All of these stages have not yet been experienced by the whole of humanity; the successive "economies" must therefore be considered as spread over time and coexisting in today's world.

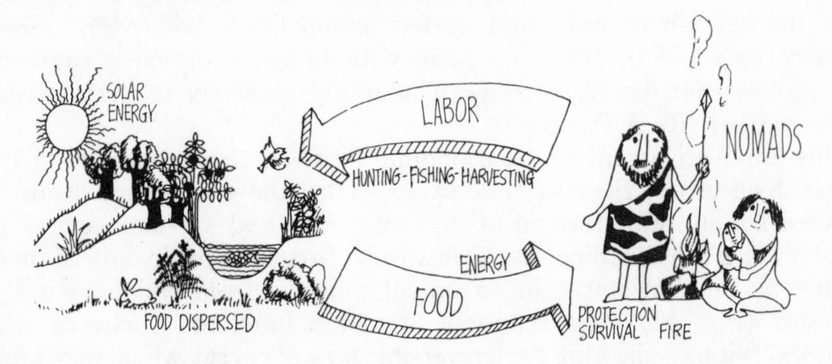

Fig. 12

The first stage is characterized by the conquest and mastery of fire. Man lives as a nomad, moving constantly in search of food and shelter. The essential function he assumes is to ensure his own survival. His main activity is to gather foods dispersed throughout his environment by hunting, fishing, and harvesting. In this way he obtains the calories that enable him to maintain his activity and assure his subsistence. Activities such as moving, fighting, and making an extended effort demand significant amounts of energy. Thus it is impossible for the nomadic hunter to maintain a sufficient reserve or "capital" of energy and skills with which he might speed his development (Fig. 12).

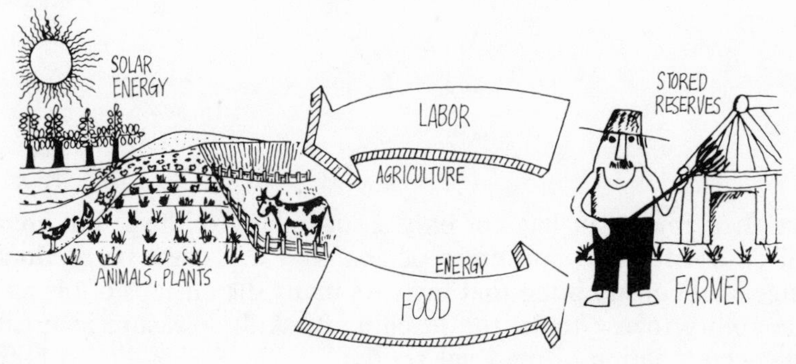

Fig. 13

The second stage comes about with the "domestication" of solar energy through the development of agriculture and the domestication of animals, both important sources of energy. This stage, which first took place about ten thousand years ago, sees man settling in sheltered and fertile zones. He can now store grain, accumulate energy, and use his reserve of energy for other activities. He now produces, thanks to solar energy, the food supply that assures his survival, and he uses animal energy to run rudimentary machines and to move him about (Fig. 13).

The third stage witnesses the appearance of more perfect tools, the concentration of work in cities, and the advent of organizations and workshops that permit the large-scale development of the work of the artisan. The quantity and diversity of objects made by artisans become sufficiently great that manufactures serve as the basis for barter. One exchanges, according to carefully specified rules, this object for food, that animal for so many items. The laws of barter insure the balance between manufactured items and consumed products; this balance is accomplished through an intermediate zone of exchange, the market (Fig. 14).

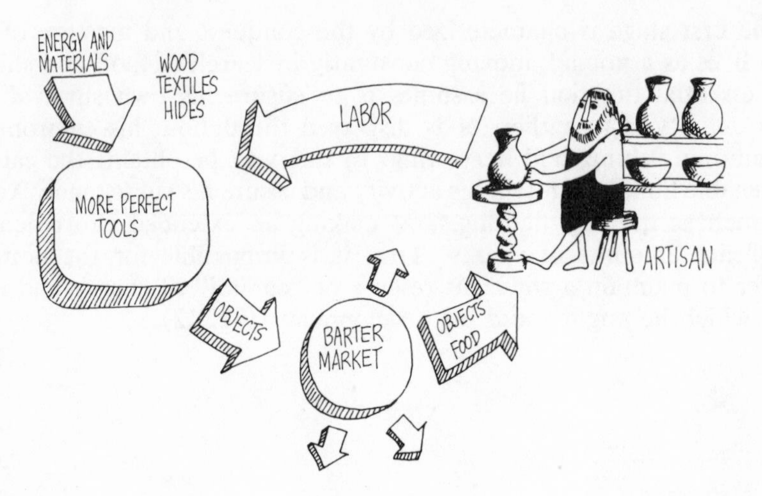

Fig. 14

Since that time, man has not only assured his livelihood but engaged himself as creator and consumer of goods in a process of production, exchange, and consumption that involves many dimensions of his nature: art, the ability to use tools, the teaching of skills, pleasure in creation, and the accumulation of material goods.

The fourth stage is the preindustrial era. The tools perfected by the artisan made possible the manufacture of simple objects for precise needs. These tools are now replaced by machines, operated by the elements, human energy, or animal energy, which lead to an acceleration of the rates of production. The density of population and the potential for exchange found in urban concentrations make possible the division of work and the lengthening of the process of production. Thus the activities of various producers become intricately linked in the interdependent chains and networks that are needed to manufacture complicated objects step by step (Fig. 15).

The use of money on a large scale and the new forms of exchange that result drastically alter the economy. Money spreads out in space and time to affect work, barter, consumption, and savings. An hour of work performed in one place can be exchanged in another place, at a different moment, for money newly earned or long saved. The two great complementary flows on which the functioning of the economic machine is based now come into play, reinforcing and balancing each other. They are the flow of energy, materials, and information, which moves in one direction, and the flow of money (resulting from barter and transaction), which moves in the opposite direction.

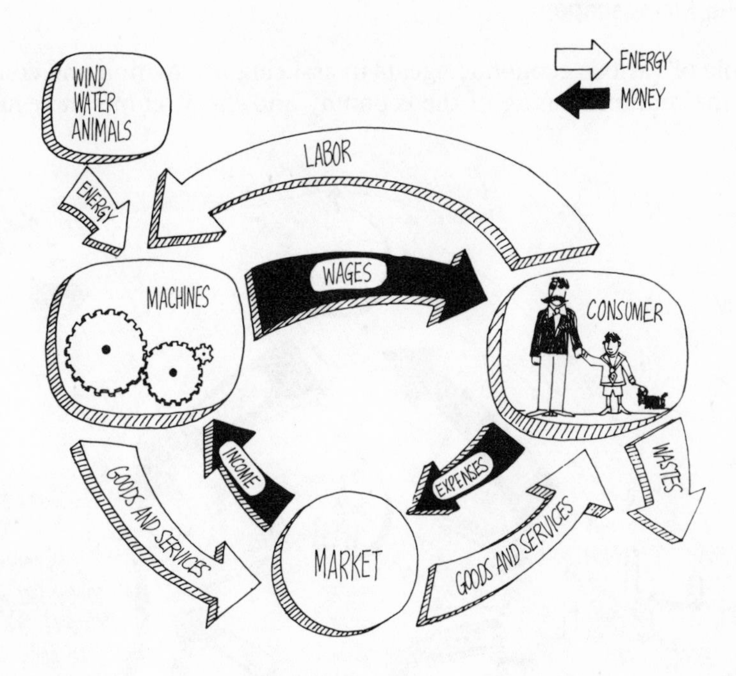

Fig. 15

The fifth stage, that of the modern industrial society, is characterized by massive use of fossil fuels (coal, oil, and gas). Other characteristics are the breaking up of work into a multitude of simple tasks which are generally without creative value, and the massive production of waste not recyclable by the ecosystem. The division of work, necessary to efficiency, requires the concentration of workers in production cells: factories and corporations.

The speeding up of the economic machine, required by economic growth, involves a growth in production and consumption. The accumulation of capital (equipment and finances) and capital knowledge (techniques and skills) has a catalytic effect on the acceleration of growth. The complexities of production require an increasing educational standard for those who would conceive, control, and serve the industrial machine.

The Economic Machine

I shall use Figure 16 to summarize the functioning of the economic machine. Its dynamics can be better understood if one follows the logic of the ecosystem: the great cycles and the main flows (energy and money), the role of various economic agents in assuring production and consumption, the malfunctioning of the economy and the attempts at regulation.

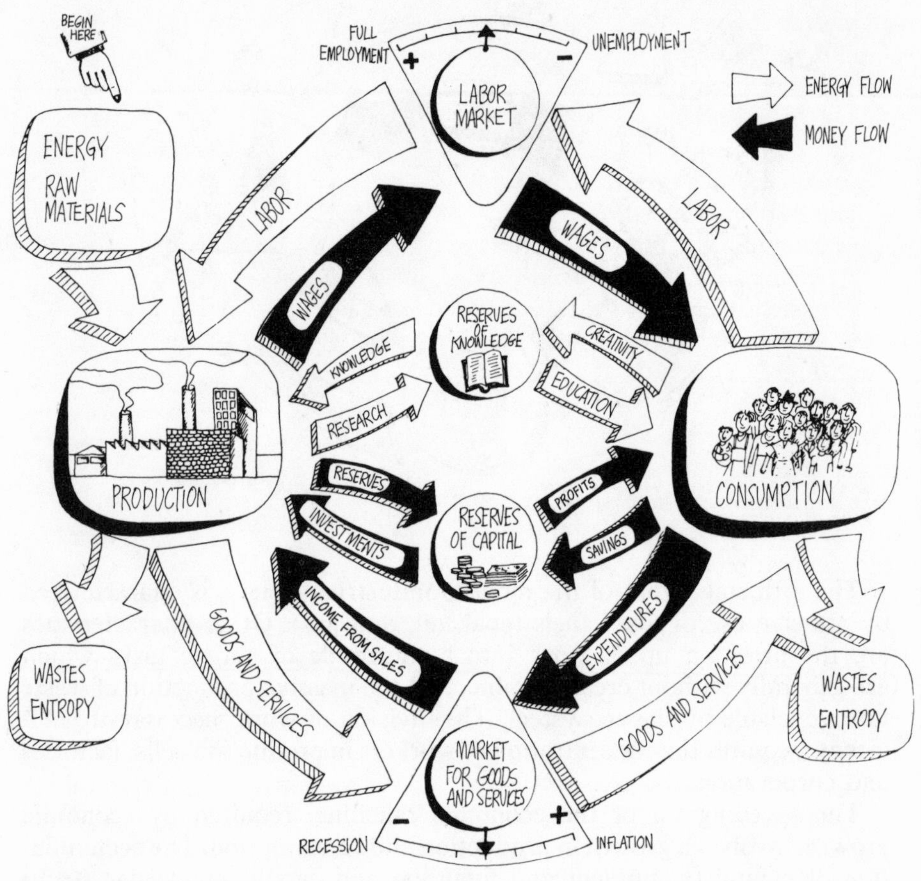

Fig. 16

Classic models consider the economic machine to be a *closed* system even though it is a system *open to the environment* and not beyond or above the rules of energy. In order to have production, energy must follow its inevitable flow from a hot source to its breaking down in irretrievable heat in a cold sink.

The diagram illustrates this expenditure of energy as it travels through the economic system. The irreversible flow enters at top left, above "production," circulates in the form of goods, services, and labor, and emerges in the form of lost heat and unrecycled wastes (entropy).

One may wonder how goods and services can constitute a flow of energy. In fact material goods—"products"—are the result of transformations involving energy, information, and raw materials. They can be considered *informed matter,* matter that has received a particular form or that has been "informed" as the result of man's activity.

Matter is condensed energy; information is a form of potential energy. Goods (including foodstuffs) and services are therefore equal to a flow of energy. To each item of goods there is attached an "energy cost," say in kilocalories (see p. 114). The feedback of energy in the form of work can be expressed easily in kilocalories expended per hour of work or in some other appropriate unit.

The sequence of white arrows, then, represents the energy that runs the economic machine. Clearly there are an expenditure of energy, as irreversible flow, and a global production of work.

Yet there is another flow related to the first: the flow of money. It runs in the opposite direction to the flow of energy. In effect, monetary units are exchanged for hours of work, information, or calories. The flow of money and the flow of energy balance each other and regulate themselves through "detectors" (cashier's desks, bank counters, transactions of all kinds) capable of measuring and balancing the speed of the flows that move in one direction or the other. Exchanges are made possible by a generally accepted system of prices and values that provides a basis for comparison and transaction. As the economists say in an almost poetic manner, the value of a good or a service is established at "the convergence of scarcity and demand." Price is the expression of this value of exchange; it is a "value meter" of distinctly practical use, since it constitutes an item of information that, while artificial, is essential in the functioning and regulation of the economic machine.

The monetary flow allows exchange by separating barter into two stages: one can sell what he possesses (his time, for example) for money, and one can buy with this money the goods and services he desires. Thus money is the lubricant—or the ballbearings—of the economic machine. Each bearing turns at the point of contact in a direction opposite to that of the flow of energy or work.

The speed of the circulation of money and the intensity of its flow depend on forces brought into play by the principal actors of the economic life, the *economic agents.* The two principal economic agents, the *producers* and the *consumers* (also called "industry" and "households"), are shown in the diagram. The other agents are *financial organizations* (banks), *the state,* and *foreign markets.* The diverse economic agents act as centers of decision that make choices and exercise powers which are then translated into forces capable of controlling, channeling, and orienting the flows of energy and money that circulate in the economic system.

Man is both producer (in industry) and consumer (in the marketplace). (In the ecosystem, remember, the two functions are accomplished by very different organisms—green plants and animals.) Yet man is much more than a simple producer or consumer.

In his role as "household," what activities in the economy can we attribute to him? His work in industry makes him a producer of goods and services. In exchange for this work he receives a salary, an income that permits him to assume the function of consumer (in the perspective of classic economic theory, to accumulate goods and services to satisfy his needs). He also has the power to save money, thereby creating capital.

Above all, man is a creator. He creates information, technology, art, new ways of living and thinking. He can even store knowledge or ideas, thereby creating "knowledge capital."

The function of producing goods and services is assured by industries (see p. 33). The relationship between the functions of production and consumption appears in Figure 16. The input is a flow of energy, raw materials (or semifinished products), work, capital, skills, and income. The output is a flow of goods and services, salaries, innovations, reserves transferred to banks to be stored, waste material, and irretrievable heat.

Following the arrows that indicate what enters and leaves "consumption," we see that input is a flow of goods and services, salaries, income to be saved, and education, and output is work, expenses, savings, new information (creations and inventions), and waste materials.

Producers and consumers can store two kinds of reserves: money (gained from past or present work), which creates capital, and information and skills, which create "knowledge capital."

Regulation of the flow of energy and money is carried out in part at the level of the labor market, in part at the level of the market for goods and services.

The three additional categories of economic agents (not shown in the diagram) have roles that are important as much for the regulation they perform as for disruptions they can introduce into the economic system.

Financial organizations—principally the banks—play a buffer role as money "reservoirs." Through the extension of credit to industries and to private individuals (credit being only another form of exchange), through investments, bank savings, and the issuing of bank notes, the banks regulate economic activity by controlling the rate of flow of money and the value of accumulated stocks. This constant adjustment of the monetary mass theoretically assures a balance between supply and demand in the market for goods and services.

The state plays an essential role in the regulation of the economic machine through budgeting and planning and by making direct purchases. Taxes and levies, subsidies, priority assignments of resources to one economic sector or another, control of the rules of competition, the fixing and freezing of prices, restrictions on credit, measures favoring exports, and the devaluation of money are examples of ways in which the state influences the economy.

Foreign markets are the rest of the world—everything located beyond the borders of the state. The state exports and imports goods and services; the difference between total exports and total imports constitutes the state's balance of payments, which has an important role in the regulation of the economic machine. Foreign markets also react to disturbances, in ways often difficult to anticipate, that influence the entire economy of a country. Political crises, devaluation or reevaluation, increases in the prices of energy and raw materials also affect the economy.

Recession and Inflation

Let us illustrate the functioning and the regulation of the economic machine by considering the acceleration and slowing down of the flow of money and energy—which are well-known symptoms of inflation and recession. There are three simple but widely used indicators that measure the effects of control as exercised by the state or financial organizations on the economic machine: prices, employment, and the balance of payments.

The relationship between the flows of energy and money can be compared roughly to the coupling of two wheels turning in opposite directions, one inside the other. The outer wheel is moved by the expenditure of energy as it breaks down in the economic machine; it turns the inner wheel by means of a series of bearings. The inner wheel can also be braked or accelerated, thereby slowing or accelerating the movement of the outer wheel. This crude model will serve to illustrate different aspects of recession and inflation (Fig. 17).

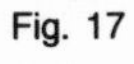

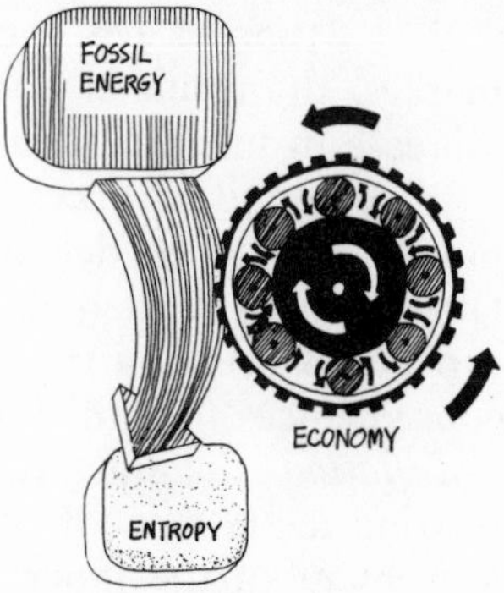

Fig. 17

Recession is characterized by the slowing down of the flow of money in relation to the flow of energy. (In terms of the preceding example, the inner wheel is braked and helps to slow down the outer one.) When the monetary supply diminishes (analogous to reducing the number of bearings), exchange becomes more difficult; friction increases and the "viscosity" of the market is raised. Locally there will be surpluses and an excess in the flow of energy. In the labor market the demand from industry will not be great enough to satisfy the higher supply, and in the market for goods and services supply will also be higher than demand (Fig. 18).

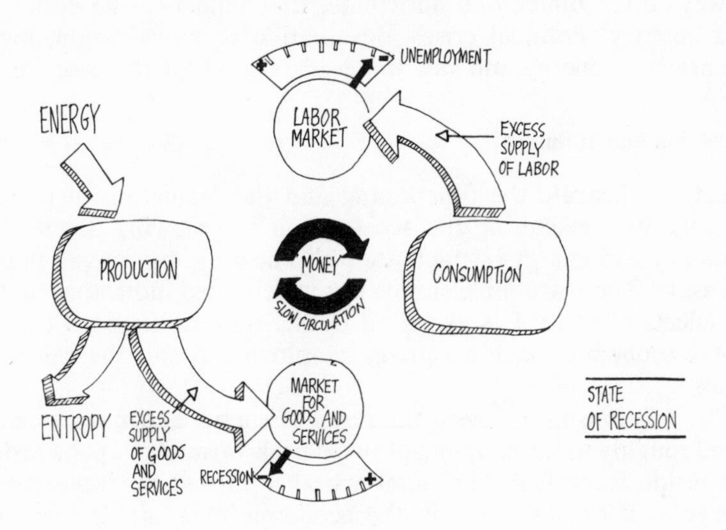

Fig. 18

When there is less money in proportion to the goods offered on the market, the result is a lowering of prices, a fall in production, and an increase in unemployment.

In a period of recession consumers prefer to wait before spending money; this lowers demand and slows exchange to an even greater extent. One becomes locked in a vicious circle—a spiral of recession that is capable of bringing the economic machine to a stop and having serious consequences for individuals and nations.

Inflation is characterized by the acceleration of the flow of money in relation to the flow of energy—by a higher "fluidity" of exchange. (The movement of the inner wheel increases the speed of the outer one, a little as though one had added bearings.) The same effect is obtained when the flow of energy slows down, which happens when energy becomes scarce and expensive (Fig. 19).

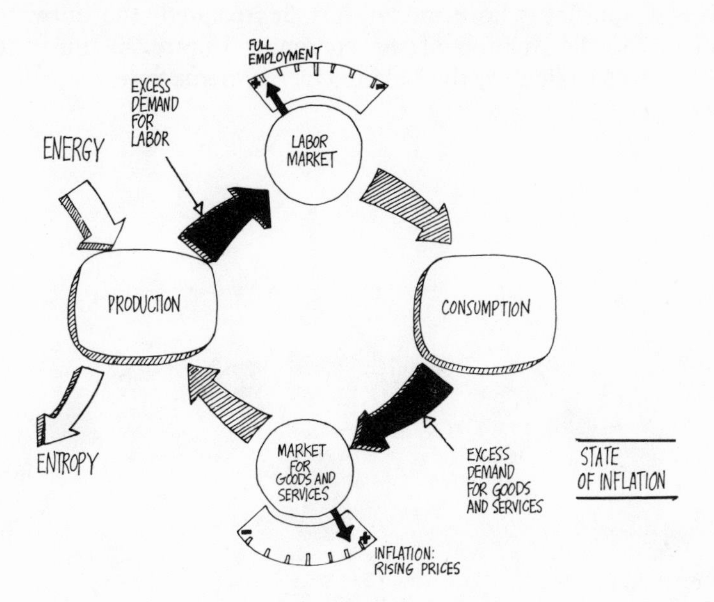

Fig. 19

The monetary supply grows, as does its rate of flow. Thus the value of money falls and buying power diminishes. Everyone wants to buy before prices go higher. Prices mount and production increases to satisfy demand, but with the increase in salaries the costs of production rise and prices go up again.

We are locked in another vicious circle—an inflationary spiral that leads not to a stop but to an uncontrollable running away of the machine that brings tensions and inequalities to the society. With the flow of money moving more quickly and affecting the flow of energy, bottlenecks are created in the labor market and in the market for goods and services. The demand of consumers is greater than the supply of producers; this causes a constant escalation of prices.

The effects of recession and inflation are obviously dangerous, although a slight inflation can be favorable to expansion and to full employment. In fact economic experience shows that there seems to exist an inverse relationship between inflation and unemployment. This poses a problem for those responsible for economic policy, for it is generally in a period of inflation that full employment is maintained.*

In the area of foreign exchange a rise in prices can slow exports and increase imports, which upsets the balance of payments. When capital leaves a country whose money has depreciated, the outward flow can lead to the devaluation of the currency. Figure 20 illustrates the role of the third indicator, the balance of payments rate.†

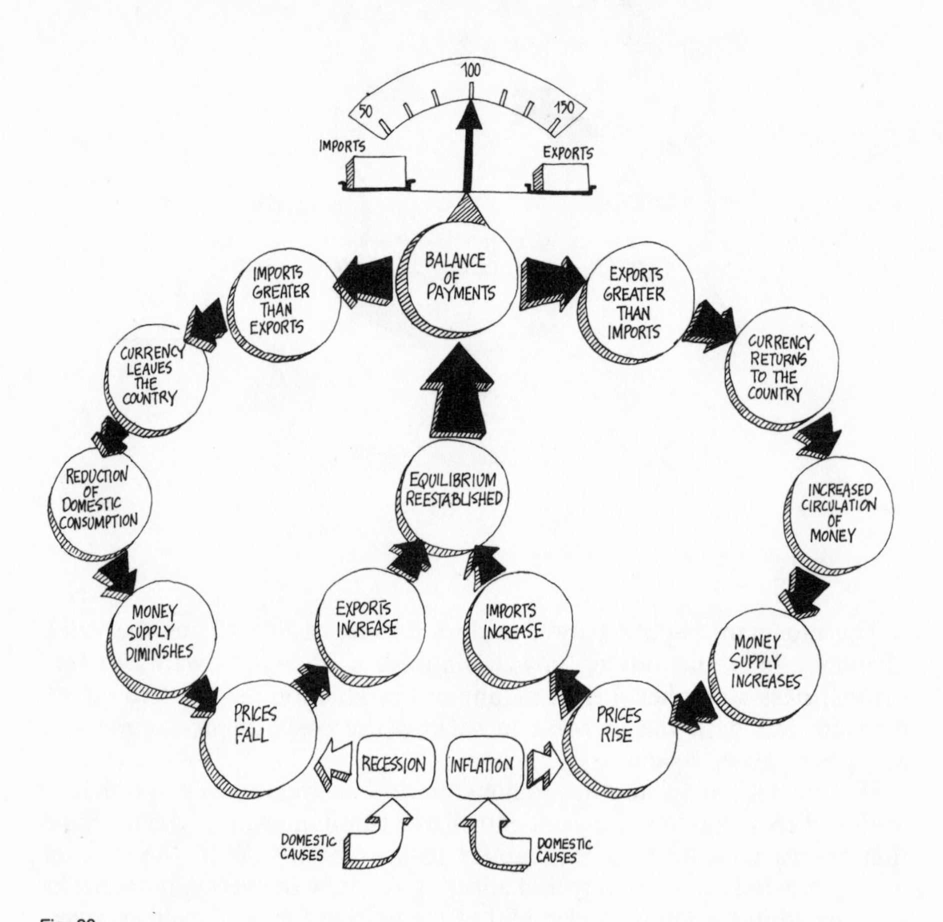

Fig. 20

---

* Except where there is *stagflation,* or inflation plus recession.

† The balance of payments rate is the relation between exports and imports taken to 100. If, for example, exports = 1 and imports = 2, the relation is $1/2 = 0.5 \times 100 = 50$, which is unfavorable for the economy of a country.

Those responsible for the economic policy of a country will try to keep the balance of payments rate at or above 100. This implies certain constraints: some inflation is maintained and, in spite of relatively elevated prices, exports must continue. On the world scale the fact that all countries would like to export more than they import leads to tensions and inequalities.

The actual situation is infinitely more complex than that we have described, and this is what makes the regulation of the economy so delicate. Ideally, in a system of free enterprise, some sort of automatic price regulation in the market would allow supply to adjust to demand. This ideal regulation is illustrated in Figure 21.

Growth in demand causes a rise in prices (goods become more scarce). Businessmen will invest and hire additional labor in order to increase production. If the supply of goods and services resulting from this increase in production exceeds the demand of consumers, prices will fall and manufacturers will cut back production.

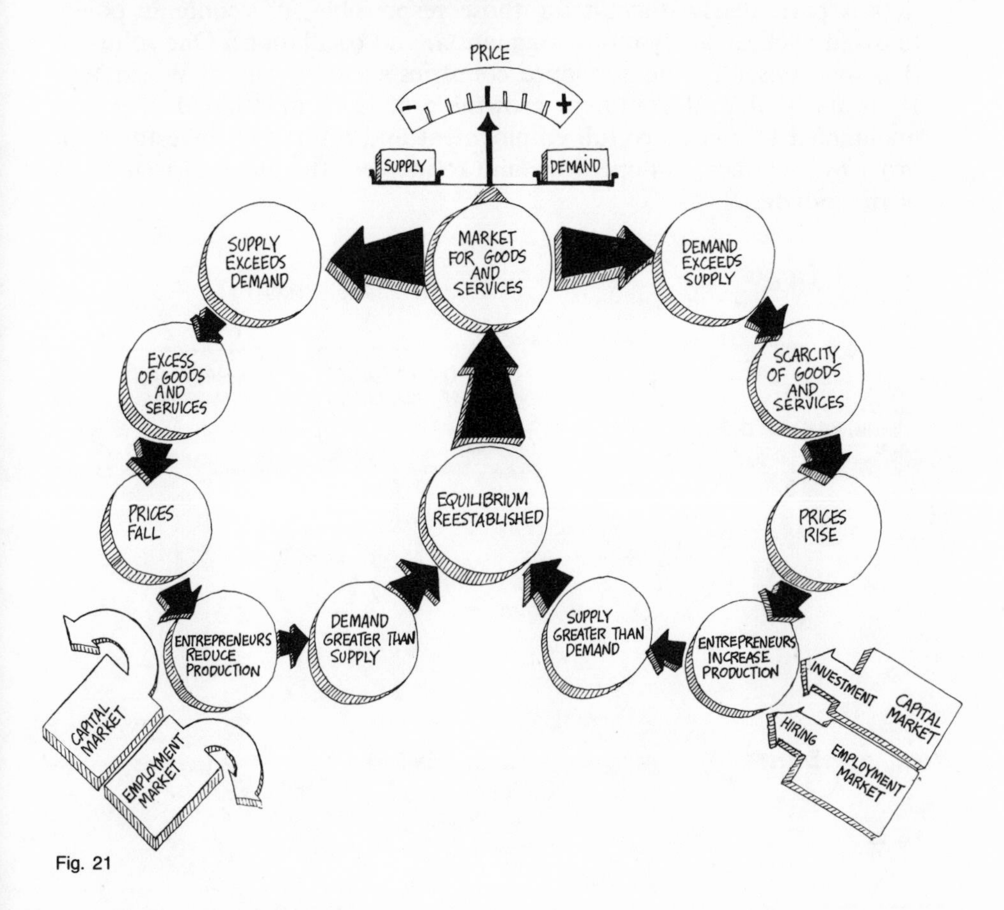

Fig. 21

Unfortunately, automatic regulation of the market by prices cannot really work. In a free system the consumer should be all-powerful, capable of exercising a permanent "right to vote" represented by his freedom to purchase or not purchase a given product, to boycott or to favor any segment of the economy. In fact, by reason of public and private investments which precede demand and guide production, by reason of the power of advertising, the monopolies created by multinational companies, and the weakness of consumer organizations, this "right to vote" does not constitute a true regulatory power. Nevertheless the whims of the consumers, panic, and a conscious desire for nonconsumption in a given sector can all create waves that can disturb the economic system.

Banks and the state act by regulating the speed of flow and the rate of saving of money. In a period of inflation or recession the state can intervene at the level of prices (price controls, price freezing), foreign exchange (tariff barriers, controls on currency, devaluation), investment (major projects, high technologies), taxes, and salaries.

It is particularly difficult for those responsible for economic policy to avoid cyclical fluctuations, stagnation, and oscillations. One solution, that was chosen by the developed countries since the end of World War II, is the policy of continual growth: a state of inflation deliberately maintained to guarantee full employment and returns on investment in order to keep factories operating and to increase the material well-being of the individual.

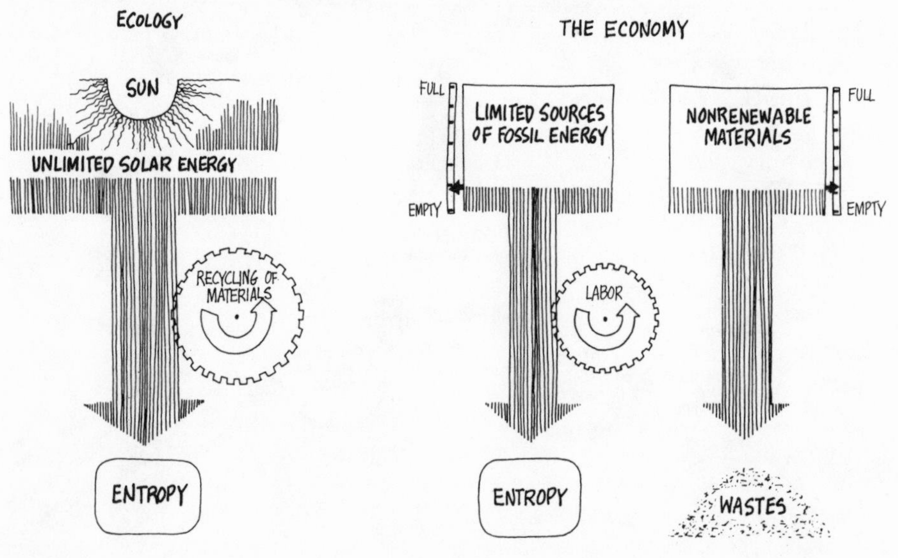

Fig. 22

But there is a price on everything. Accelerating the economic machine means pumping more energy from a depleting reservoir and dumping more waste and heat into the environment. Here lies the fundamental difference between the ecological machine and the economic machine. The basis of ecology is an irreversible flow of solar energy in *unlimited* quantity and a *permanent recycling* of materials; the basis of the economy is an irreversible flow of fossil energy from a *limited* source and an irreversible flow of materials from a *nonrenewable* reservoir of resources (Fig. 22).

3. THE CITY

Today the laws of the economy and ecology come face to face in a type of organization that is new in the history of the ecosystem. It is the nerve center of an immense network of exchange and communication, one of the most complex organizational forms in the social fabric: the city. Already more than 50 percent of the world's population live and die there; in the year 2000 about 80 percent of mankind will work and live in cities of more than 100,000 inhabitants. The city is born, develops, diversifies, and dies. It transforms energy, shelters man, and facilitates communication.

For millions of men and women the city is the principal place of employment. The development of business and industry has conditioned the growth of the city and the city in turn has modified the structure of industry. In their reciprocal adjustments and the special conditions they have created for labor and commerce, city and industry have brought about new ways of life and new aspirations. At the same time they have imposed between man and nature a sort of external biological layer that sometimes oppresses and often isolates us.

The city was born out of the needs of man: the physiological and utilitarian needs for shelter, food, health, communication, trade; the psychological needs for esteem, respect, education, and power.

The structure of the city acts as a catalyst to accelerate the development of philosophical and religious ideas, science and technology, the arts and political concepts. Through the organization of expansion, confrontation, experience—and restraint—this prodigious center of innovation attracts, promotes, and engulfs both men and ideas like a whirlwind.

The city is a communications machine, a massive network whose principal activity is the acquisition, processing, and exchange of information. It promotes in particular the plurality and variety of communication and exchange: for employers who can use a wide range of talents and specialties; for employees and consumers whose abilities and whose demands for goods and services are highly diversified.

The combination of these factors in the heart of the city has contributed to the almost explosive development of great metropolises during the course of this century. In 1850 only four cities in the world had more than one million inhabitants; in 1900 there were twenty; fifty years later, 140.

Each breakthrough in one field has repercussions in another that encourage its own development: scientific discoveries, industrial productivity, new products, new means of communication, new methods of transportation. At the same time there arise new ways of living, demands, constraints, conflicts, social readjustments. Thus the organic complexity of cities is woven.

The first cities were born nearly five thousand years ago in villages located in fertile areas that fostered communication: the fertile crescent of Mesopotamia, the valleys of the Nile, the Indus, and the Yellow rivers. The fertility of the surrounding land permitted the use of solar energy in agriculture. The storing of food and energy made possible the maintenance of the complex structure of the first cities, while the production of surplus energy increased the rate of growth and development. Communication in the deltas and by means of large river systems developed exchange, barter, and trade, allowed the confrontation of cultures, and inspired technical and social innovation.

Before 1850 there was no urban society; the great majority of people lived in villages where they produced for themselves everything they needed. From raw materials and energy (mainly food, fuels, wood, textiles, leather) they produced goods and services useful to the community. Thus the village was able to assure its own maintenance and survival.

The first large cities were the home of the leaders and representatives of the society, the statesmen, clergy, military, nobles, bourgeois, and great merchants who formed a minority of about 20 percent of the population. This elite survived until the end of the eighteenth century thanks to their resourcefulness and the contributions of energy in the form of the labor of the villagers and the collection of taxes and levies of all kinds.

During the nineteenth century and early in the twentieth the Industrial Revolution and the division of labor led to specialization. Long-distance communication systems (train and telegraph) combined and reinforced each other, attracting to the cities an ever-increasing flow of the population.

The autocatalytic effect characteristic of large metropolitan areas began to exert its force. The pulling power of the cities' freedom of choice, higher wage levels, and possibilities for amusement and success accelerated the drain of people, energy, and materials from the periphery of the cities.

The modern metropolis rose from the density of population, the horizontal and vertical advance of construction, the organization of means of communication (automobile, telephone, elevators), and the creation of rules and codes that allowed cities to control their main functions (regulations governing working hours and traffic control, for example). At the same time cyclical waves were created by the great daily migration of workers into and out of the center city. The city has become a gigantic pump that sucks in and pours out, certain sections alternately filling and emptying in accord with working hours and weekends.

The growth of the city and its diseases, the multiplicity of its functions, and its daily behavior all suggest that the city reacts like a living organism interacting closely with an environment that it influences indirectly and that transforms it in return. Like the coral reef, the beehive, and the termite colony, the city is at the same time the support and the consequence of the social organism which lives at its heart. It is particularly difficult, if not impossible, to separate within every organism structure and function; thus one must not fear the analogy between the city and a "living organism" as long as the term appears within quotation marks.

What are the principal elements that make up a city, and how do they interact? On the map of a city the structural features hide the functional processes. Among the streets, avenues, and blocks of houses one sees stations, monuments, hospitals, and administrative buildings, yet everything seems frozen. The functional aspects of the city—its dynamics—escape observation. To grasp their complexity one needs the equivalent of an atlas that brings out in sufficient detail the urban body tissue area by area, showing the flow of energy, material, and information that circulates among businesses, administrative centers, residential zones, the environment. By regrouping certain main functional categories of the urban tissue, it appears possible to glimpse the overall portrait of the city and even to compare some of its structures and functions with those of other organisms, whatever their level of complexity (Fig. 23).

The interaction between individuals and organizations through communications networks makes possible the major functions of the urban system: utilization of energy and elimination of waste; production, consumption, and administration; culture, leisure, and information; communication and transportation; shelter and protection. The various functions result from different structures.

Housing. The largest single area of cities is occupied by dwelling places, which assure the establishment and the protection of the family unit. Residential zones represent, on the average, about 40 percent of the area of all cities.

Business and commerce. The city is the place of work for the majority of city and suburban dwellers, and industry produces the goods and

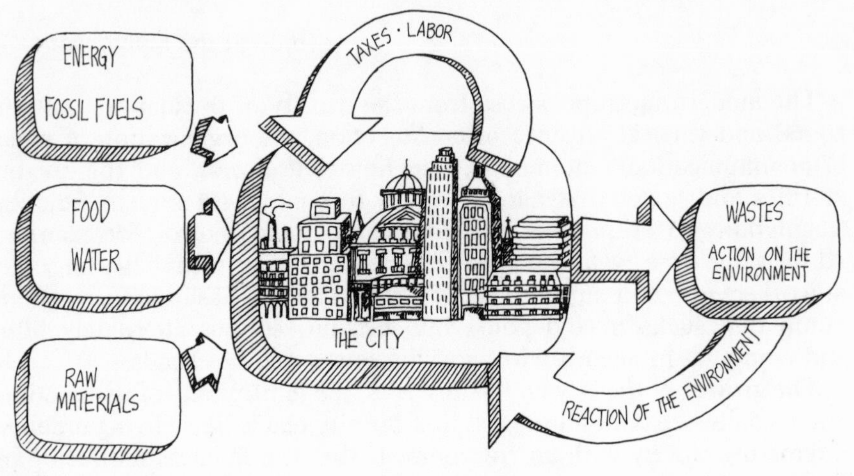

Fig. 23

services needed by the community. The distribution of products is carried out by the commercial sector, from its small shops to its chain stores and supermarkets. Food accounts for 25 percent of the budget of consumers in industrial countries.

Communication and transportation. These networks differ according to whether they transport people, materials, or information. The first two instances involve a layout of avenues and streets, a network of urban transport (subways, buses, taxis), and intercity and international systems (railway stations, seaports, airports). The third instance involves telephone lines and cables, telephone and postal systems, radio and television stations, the press and publishing houses.

Reserves. The principal storage areas in cities are distinguished by the character of what they store—energy, materials, or information. Large storage tanks hold energy in the form of fuel oil, gasoline, and natural gas. For perishable foodstuffs there are markets, slaughterhouses, refrigerated warehouses, grain silos. All kinds of materials are kept in stores and warehouses, and drinking water is kept in cisterns and huge reservoirs. Information is stored in libraries, archives, and computer banks; money is kept in bank accounts and vaults.

Administration and finance. Another important area of large cities is occupied by those agencies that contribute to the regulation of the social and economic balance: ministries, local and national governments, banks and other financial centers.

Distribution of energy and elimination of waste. Energy enters and moves through the city by means of electricity systems, gas mains, and gasoline stations. It leaves in the form of heat and garbage collected in sewer systems or by trucks. Wastes are eliminated in part in purification plants and incinerators, or they are accumulated in garbage dumps or used as landfill.

Other types of centers are more directly related to the daily activities of the inhabitants: culture and artistic life (museums, monuments, theaters); health care (hospitals and clinics); education (schools and universities); leisure activity and amusement (cinemas, stadiums, playgrounds, parks, cabarets); protection and security (fire and police stations, military installations, prisons); religion (churches, cemeteries). These different components of the city are often grouped into distinct sections within the city: business, amusement, and university districts, industrial or commercial zones, government buildings, museums, and green areas that are the "lungs" of the city.

Every city has its history; every city also has its daily routine. It feeds on tons of foodstuffs, fuels, and water to support the activities of its citizens at home and at work. For a city of one million inhabitants, daily consumption amounts to about 2,000 tons of food, 4,000 tons of fuel, and 630,000 tons of water.

The city continually absorbs materials that replace worn-out structures or are used in the construction of new ones. Like a living organism, the city is the seat of a perpetual turnover of all its elements. This dynamic renewal can be seen in the coexistence of junk yards, sometimes a city block in size, with building supply yards. Such turnover has an effect at all levels of the organization of the city.

All cities release into the environment their metabolic wastes. For a city of one million inhabitants, daily wastes amount to 500,000 tons of used water containing 120 tons of solid particles, 2,000 tons of garbage, and 950 tons of atmospheric pollutants. The effect of pollutants on the lives of the citizens is only too well known, yet one such effect is worth discussing because it is a direct result of the metabolism of cities: the modification of their microclimate.

The city is a source of heat as a consequence of man's activities (heating, air conditioning, factories, automobiles). It also creates a "heat trap" because vertical surfaces reflect and amplify solar radiation, because the irregular outline of the buildings increases turbulence and reduces the escape of heat, and because running water from precipitation is immediately collected and drained off and cannot contribute to the cooling of walls and soil through evaporation. The temperature of the city, then, is always several degrees higher than that of the surrounding countryside. To the effects of this blister of heat add the effects of dust and aerosols in suspension in the air. These create condensation nuclei, causing the habitual haze and clouds that so often obscure the skies over large cities. The result: 30 percent fewer sunny days in winter and 10 percent more precipitation each year than in the immediate environs of the city (Fig. 24).

Fig. 24

Another manifestation of the daily life of the city is the movement of its workers. In most large centers of population this creates a succession of concentric circles of residences traversed by busy roads. Once a centripetal movement has drawn the hordes into the city, automatic regulation comes into play: noise, pollution, stress, the high cost of living, and a lack of security cause a centrifugal movement toward the green suburbs and the country. In some cities there are downtown areas where glass buildings and slums stand side by side, busy with people in daytime and deserted at night, when violence and fear reign; the population live in the suburbs, spending an hour or two each day commuting to and from work by car or train (Fig. 25).

Fig. 25

The city appears to be a self-regulating system that controls and balances the flow of its people between its center and its periphery. In the course of history the city has passed through a phase of explosive growth, followed by a period of stabilization, then stagnation marked by the degeneration of some areas, the appearance of slums, the further exodus to the suburbs, and the erosion of the center city.

4. BUSINESS AND INDUSTRY

Following the Industrial Revolution business concentrated in the cities, where it found ideal conditions for development: density of population, abundant manpower, intensity of exchange (of goods and services, money, information), and trade and commerce. The business of a city determines its personality, its shape, and its appearance. In almost all great cities there exists one or more functions of production that characterize those cities and give them life and purpose. The excavation of ore and the production of energy, large textile factories, foods, chemicals, iron and steel, mechanical engineering, and electronics are examples of such industries.

Today the businesses located in the centers of the great cities are chiefly producers of services, firms oriented toward commerce and distribution, or the central offices of national and international corporations, which combine administrative, commercial, and financial activities. Factories are leaving the central areas of the large cities to relocate in the suburbs or in the country, where they are closer to the sources of energy, manpower, and raw materials.

To a stranger to the business life all businesses seem the same. The visitor sees only offices with the usual instruments of communication: telephones, typewriters, copying machines—occasionally workshops and laboratories. Their internal structure discloses itself on organization charts, but the charts do not show the movements of men and information that are the true activity of the company.

Nevertheless each business has its own life. It is born, it grows, develops, reaches maturity, and dies. Each business is one cell of production in the social organism; together a country's industries constitute a megamachine of production. Like a pump of gigantic dimensions, it sets in motion the flows of energy and money that course through the veins of the economic system.*

Business brings together various economic factors, organizes them, and uses them to produce goods and services that can be marketed. A business can be represented by a single person (a lawyer or an artist, for example), or it can take the form of an agricultural enterprise or an organization of craftsmen. In this more general view, "*business* is any activity that ends in the selling of a product or a service in the markets for consumption or production of goods" (Albertini).

Business is also a decision center capable of providing its own autonomous economic strategy—one whose principal objective must be to "max-

* The following describes in general terms the functioning of a classic type of enterprise, that of the "growth economy." Once again the discussion will focus on those characteristics that are common to all complex systems.

imize profit within the technical and financial constraints that enclose it" (Attali and Guillaume). Thus business exercises two principal functions, one at the individual level, the other at the level of the society. The first function is to produce goods and services that will satisfy the needs of men; the second involves creating wealth, or generating through growth a surplus monetary value that, reinvested in part in the economic system, contributes to raising the standard of living of the entire population.

What makes a business run? The first requirement is organization, the establishment of specialized departments within the company and networks of communication that link them together. The production department brings together the factories, workshops, and machines. The commercial department oversees the system of distribution. Administration and management are the organs of planning and control. Research and development are the source of new products.

The second requirement of a business are the factors of production, the elements that actually set the company in motion: labor, capital, energy, material, and information.

Labor is the energy provided by the workers, the employees, and the officers of the company, who manufacture products and gather and process information.

Capital is financial resources and equipment.

Energy and *material* are the fossil fuels, electricity, steam power, primary raw materials, and semifinished products that will be used as starting materials in manufacturing or assembly.

Information is technology, licenses, patents—the intangible assets that are the result of the experience of the members of the company and their accumulated knowledge.

The material goods produced by the company through the combination of these factors are intended either for other businesses (production goods) or for individuals (consumer goods). Nonmaterial goods produced by the company are services (transportation, advertising, consultation, insurance). The company purchases its factors of production in specialized markets. This is shown in Figure 26, which is based on the diagram on page 25 but which opens up the feedback loops to illustrate the application to business and to show inputs and outputs.

Business buys or rents in various markets the factors necessary for the production of goods and services. When it needs money to develop or maintain itself, it can "rent" (borrow) money from banks by taking short-term or long-term loans. It can also "buy" money by paying the sellers in money of a kind peculiar to business: stocks. The sellers take a share (representing a fraction of the property) of the company and become stockholders.

Business deducts from its revenue the sums necessary to remunerate its production factors. These sums represent salaries (payment in exchange for labor), interest and dividends (in exchange for loans or capital), and royalties (in exchange for technology and patents). Money is also deducted to pay taxes imposed by the government. Business creates wealth only if it produces more in value than it consumes.

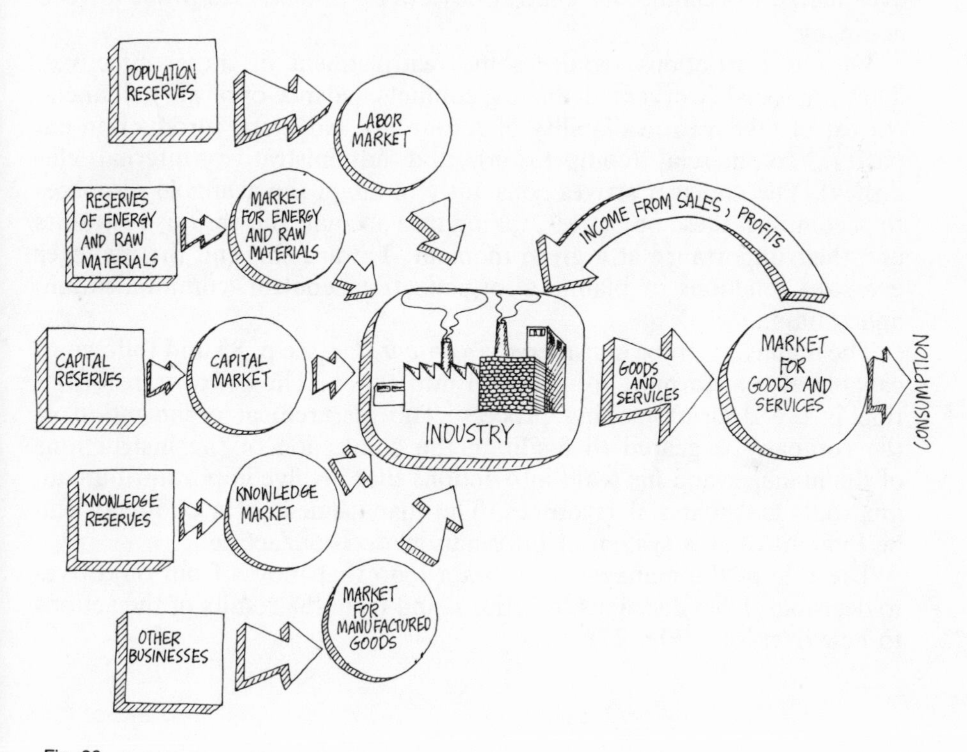

Fig. 26

The choice of a company's objectives, its methods for achieving those objectives, and the controls that will keep the company moving in the right direction are the responsibility of management, led by the chief executive. Good management—efficient direction of the company—involves the adjustment of the company's objectives according to the various limitations of the environment in which it exists.

What are the main objectives of a business? The first is improvement in production—growth in the quantity of goods and services produced. Then comes the choice of financial resources and investments that will affect the firm's potential for profit. The company must maintain its competitive position through marketing and research and development so that the demand for existing and new products will grow. The training of factory workers, office workers, and management is important. Finally, the company's social function, its role as an agent in the transformation of society, confers on it a public responsibility. And all operational objectives merge to become one central objective: to maximize profit for the company.

Various restrictions require some readjustment of these objectives. They are social (workers' demands, conflicts, balance of power), financial (threat of takeover, availability of resources), industrial (production capacity), commercial (competition), and administrative (internal efficiency). The manager strives constantly to adapt the available resources to accomplish these objectives, taking into account the various restraints and their importance at a given moment. Toward this end the manager exercises functions of planning, organization, control, communication, and training.

The manager can be considered a *comparator* (see p. 85 and following) capable of transforming information into action. The act of transformation is the decision-making process. The hierarchical organization of the company is geared to facilitate the conversion of the instructions of the manager and his team into actions that involve important human, financial, and material resources. The management of a company can be thought of as a system of information/decision/action.

The role of the manager inscribes a loop that moves from objectives to decisions, from decisions to actions, and from the results of the actions to new decisions (Fig. 27).

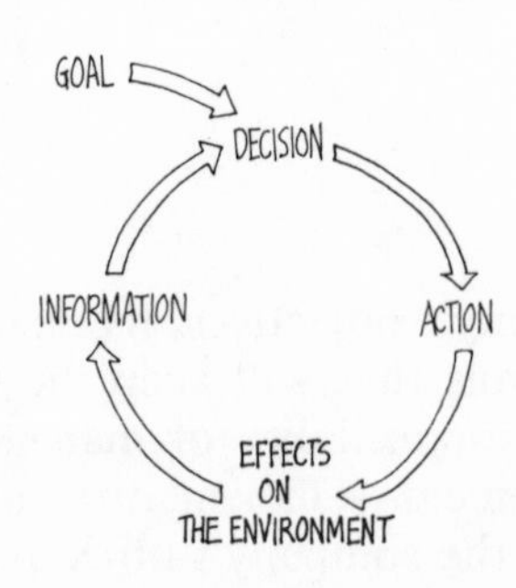

Fig. 27

This role necessitates two modes of action that may appear to be contradictory. On the one hand, the manager must act as a stabilizer: to assure the survival of the company and the security of employment, he must maintain its equilibrium. On the other hand, he must assure the continuous growth of his company. The application of these two modes of action determines the dynamic behavior of the company. Like any living organism, the company can enter phases of growth, stagnation, regression, and fluctuation.

The main purpose of the long-range strategy of a company is to assure growth while maintaining balance and stability. The traditional social and financial restrictions on industry give way to new restrictions that arise from the rapid development of industrial societies and the acceleration of economic growth.

The management team must see far ahead, decide more quickly, prepare forecasts and detailed plans for development, and come up with strict methods of control. The team must also take into account changes in the environment, in technology, in the tastes and the needs of people. It must consider its competition and the general rate of expansion of the national and world economies. These demands have led companies to adopt a growth strategy founded as much on the creation of new methods (technological, industrial, and commercial) as on joint ventures with other firms or the acquisition of companies that offer perspectives of diversification.

This growth strategy leads to a financial strategy that involves selecting the kinds of resources that will maintain a rate of growth compatible with the size and objectives of the company—while allowing management or the majority shareholders to retain some freedom of action.

Reinvestment in the company of a share of the profits promotes growth and financial independence. The company that can guarantee this kind of financing releases and maintains its own explosive catalytic process. The profits earned by a well-managed company can be compared to an energy surplus; the reinjection of this surplus in the form of strategic investments (consolidation of the financial status, strengthening of work forces, development of the distribution or production networks) is a kind of self-financing. It assures at the same time the maintenance of the structure, the accumulation of capital, and continued growth. The objective of every entrepreneur who establishes a business is to achieve a continuing process of reinvestment (Fig. 28).

Even should it begin with only two persons, a well-managed company can reach a level of efficiency that will assure its maintenance and growth. Usually it takes several years for a new company to reach respectable proportions; to grow more rapidly almost always means investing more money. To "put in orbit" a new company within a very short time requires—in addition to ideas, men, and technology—a particular type of capital, "venture capital," which represents the "potential energy" needed to fuel the "reaction." The risk can be high: how much energy will have to be spent before the "reaction" stops consuming energy and begins to produce the small excess quantity that will trigger the chain reaction? On the evaluation of such a risk is founded the art of the creation of new business.

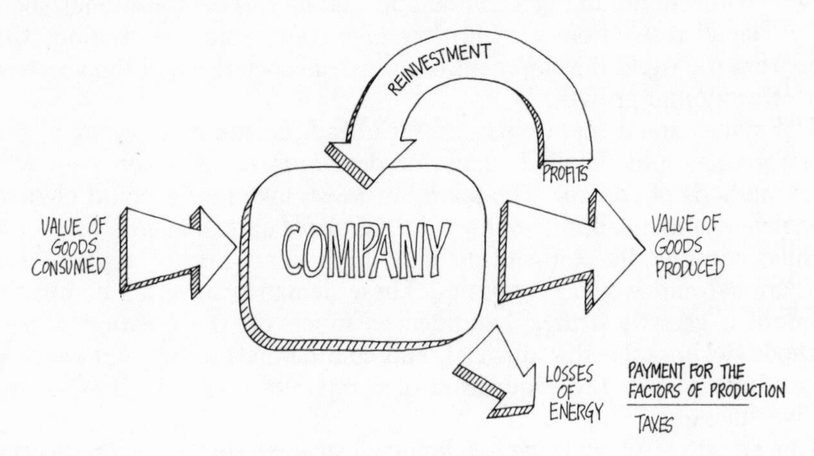

Fig. 28

5. THE LIVING ORGANISM

From the end of the eighteenth century and into the nineteenth, following progress in anatomy, physiology, and medicine, numerous naturalists and philosophers (Worms, Spencer, Bonnet, Saint-Simon) extended their concept of the organism to society as a whole (the political and social organism). Often using naive or daring analogies, some of which we now find amusing, they nevertheless contributed to broadening the horizon of our knowledge about the life of man in society.

The metaphor of the living organism actually has great evocative power. In the words of Judith Schlanger, it permits the "integration of knowledge and meaning." It embraces complexity and interdependence in an integrated, autonomous whole in which the intricacy and variety of relationships between the elements often appear to be more important than the elements themselves.

In this section and the next we shall encounter again, at the level of organism and cell, familiar principles and patterns of operation. As in our studies of ecology and the economy, we shall observe the organism and the cell through the macroscope in order to concentrate on the broad lines of their function and regulation and to emphasize fundamental ideas of a new method of approach to complexity.*

Consider a man at his place of work: he carries out a specialized function. His labor may be manual (moving or positioning objects—exerting strength) or intellectual (screening, classifying, processing information—organizing and controlling). His action on his immediate environment is thus translated into *energy* or *information* that is transmitted to other men and machines.

What does this man need to perform his work and produce his efforts? Above all he must have energy and information. Energy comes from the foods he buys and consumes regularly. Information falls into two categories: his initial capital, the education or training that gives him expertise in his profession; then the instructions that guide his work and the signals that come from his environment and from within himself. In exchange for his labor, this man receives a salary that enables him to obtain food and the other goods and services he needs (Fig. 29).

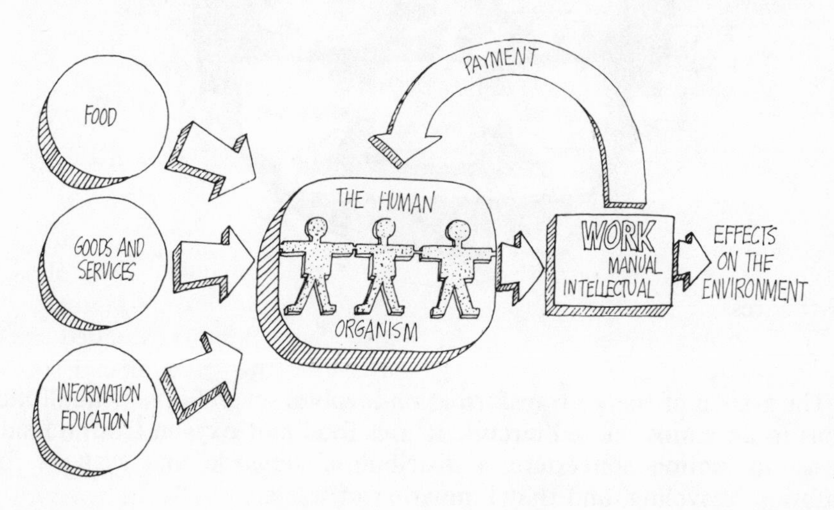

Fig. 29

* Reproduction that has no permanent effect is not considered here. The description that follows purposely uses very general terms in order to emphasize analogies between the organism and the other systems that have been described.

In order to sustain life, perform work, and receive and generate information, a particular kind of organization is needed. This organization relies on transformation centers (organs) and networks of distribution of energy and communication. The fundamentals of this organization are shown in Figure 30.

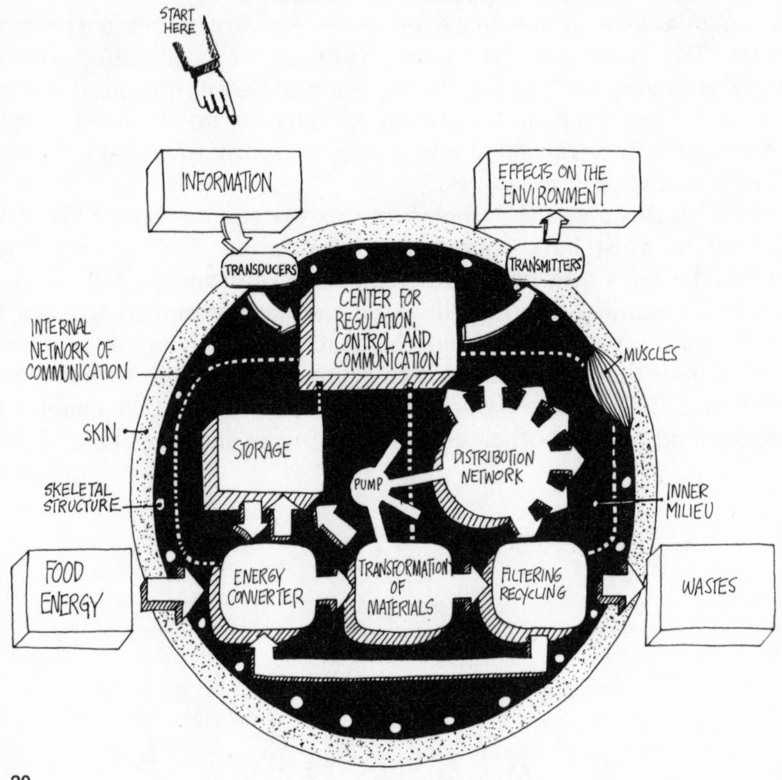

Fig. 30

The system of energy transformation involves several organs and functions in an almost closed circuit. It uses food and oxygen from outside to set in motion converters, a distribution network, and systems for filtration, recycling, and the elimination of waste.

Energy-rich foods (sugars and fats) and indispensable raw materials (proteins and amino acids) pass through the series of converters of the digestive system (stomach and intestines). In the course of various transformations the substances extracted from raw foods are either used immediately or stored for later use. Oxygen in the air is breathed in by the lungs, which reject by expiration the body's most important combustion gas, carbon dioxide. Oxygen, the energy derived from food, and other essential substances are distributed in a fluid (the blood) that circulates in a complex network. This circulation is kept up by the work of a pump (the heart) capable of pumping from 5,000 to 6,000 liters of blood per day. Because metabolic wastes and combustion gases are returned to the blood, a system of filtration, recycling, and waste elimination is needed to cleanse this vital fluid. The principal filters are the lungs, kidneys, and liver.

Blood is regenerated in the lungs through the elimination of carbon dioxide and the absorption of oxygen by the hemoglobin of the red cells. The kidneys filter and recycle blood after cleansing it of wastes; 99 percent of the fluid that flows through the kidneys is returned to the bloodstream while the remainder becomes the urine that carries off the waste. The liver acts like a chemical filter, retaining and destroying any substances that would be toxic to the system.

The information-processing system is composed of transducers and memories; organs of processing, control, and regulation; and two interconnected communications networks, one electrochemical (the nerves), the other chemical (the hormones).

The transducers transform signals from the environment into recognizable bits of information. The transducers are photoelectric (detection of light and images), acoustical (detection of sounds), chemical (detection of odors), and mechanical (detection of touch); they constitute the sensory system.

Information is stored in the memory and treated in different areas of the spinal cord and the brain—the olfactory, visual, and auditory zones. The control and regulation of major functions of the body are assumed by the brain or directly by the endocrine glands. Regulation often requires the cooperation of several organs; an internal network of signals is therefore essential. This network, by nature electrochemical, permits the transmission of an electrical impulse (representing information) through the medium of the nerves. The network also has a chemical nature: the release by an endocrine gland of a molecular signal—a hormone—in the bloodstream. All organs through which the blood flows will receive this hormone, but because the instruction that the hormone contains is coded, only the organs concerned will be instructed to undertake the regulatory action. These networks of communication are the nervous system and the endocrine system.

The body, limited by the skin, resembles a watertight bag that is 60 percent filled with water. Because the organs and their networks of communication do not have a consistency sufficiently rigid to prevent the entire body from collapsing under its own weight, the skeleton acts as a framework. Many of the 206 bones that make up the skeleton act as levers and are essential to all motion and to every movement. The contraction of six hundred muscles of the muscular system provides the motor force that acts on the levers or on tissues to bring about motion and movement.

The skin is a barrier that prevents microbes and foreign matter from invading the body. In the case of a lesion, the skin repairs itself through the process of healing. The skin's sensitive surface is capable of detecting, through nerve endings, information from the environment. It also plays a part in controlling the body's temperature. The body possesses a defense system that protects it from attack by foreign substances. Its weapons are antibodies, which are capable of recognizing and destroying foreign protein, and the white cells, which absorb and neutralize bacteria dangerous to the body.

The organization of the body enables man to act on his environment and to respond to the information or aggression that he finds in it. Physiologists have shown that the reactions of man and animals to these aggressions lead to three basic behaviors: flight, conflict, and adaptation.

When the environment becomes disagreeable, hostile, or dangerous, the organism can respond by leaving; it simply continues to change its environment until it finds a milieu in which it is comfortable. It can also attack or defend itself. And by conscious action it can modify an environment that threatens it and thus restore favorable conditions.

The body appears to be able to adjust itself continuously to new circumstances. In fact this adjustment is never perfect. Man experiences difficulties in adapting fully to a given environment; the adjustment often provokes frustration, anxiety, and illness. However, these are sometimes positive factors that are the basis of conscious or unconscious moves that lead to change or transformation.

A man threatened by the environment (or informed of an approaching pleasure or danger) prepares for action. His body mobilizes reserves of energy and produces certain hormones such as adrenalin, which prepare him for conflict or flight. This mobilization can be seen in familiar physiological reactions. In the presence of emotion, danger, or physical effort the heart beats faster and respiration quickens. The face turns red or pales and the body perspires. The individual may experience shortness of breath, cold sweats, shivering, trembling legs. These physiological manifestations reflect the efforts of the body to maintain its internal equilibrium. Action can be voluntary—to drink when one is thirsty, to eat when hungry, to put on clothing when cold, to open a window when one is too warm—or involuntary—shivering, sweating.

The internal equilibrium of the body, the ultimate gauge of its proper functioning, involves the maintenance of a constant rate of concentration in the blood of certain molecules and ions that are essential to life and the maintenance at specified levels of other physical parameters such as temperature. This is accomplished in spite of modifications of the environment.

This extraordinary property of the body has intrigued many physiologists. In 1865 Claude Bernard noticed, in his *Introduction to Experimental Medicine,* that the "constancy of the internal milieu was the essential condition to a free life." But it was necessary to find a concept that would make it possible to link together the mechanisms that effected the regulation of the body. The credit for this concept goes to the American physiologist Walter Cannon. In 1932, impressed by "the wisdom of the body" capable of guaranteeing with such efficiency the control of the physiological equilibrium, Cannon coined the word *homeostasis* from two Greek words meaning to remain the same. Since then the concept of homeostasy has had a central position in the field of cybernetics.*

The "internal milieu" is properly identified with the principal fluid that circulates through the body and washes the organs and the cells: blood plasma. Plasma is an aqueous milieu in equilibrium with the extracellular fluid found between capillaries and cells. It is a vestige of the primitive ocean inhabited by the first living organisms. Plasma accounts for 55 percent of the blood (the other 45 percent consists of red cells, white cells, and platelets). Plasma is 92 percent water and 8 percent molecules essential to life (glucose, amino acids, fatty acids, hormones such as insulin, adrenalin, and aldosterone) and ions such as calcium or sodium.

What are the main properties of plasma that make regulation effective? Temperature, maintained in the neighborhood of 37° Centigrade in man and most mammals; the concentration of calcium and sodium ions; the concentration of hormones and glucose; the pressure and volume of the blood; the number of red cells; and the acidity and concentration of water in the plasma.

Regulation is achieved by means of a control mechanism containing a *detector,* a *comparator,* and a *memory bank* that records the limits that cannot be exceeded. Each molecule or ion present in the plasma comes from a "source," can be stored in a "reservoir," and disappears in a "sink." A general model of a typical physiological regulation is presented in Figure 31.

---

* There will be more on this subject in the next chapter.

Consider, for example, the regulation of the concentration of calcium. Calcium plays an important role in muscular contraction and in the formation and composition of bones. Its concentration in the plasma is maintained in a remarkable way at a level of between 8.5 and 10.5 milligrams per hundred milliliters. Calcium enters the body daily in food (milk in particular contains a considerable amount of calcium). It can be stored in the large calcium reservoir of the bones. Very little calcium is excreted in the urine. The regulation of the concentration of calcium in the plasma is shown in Figure 32.

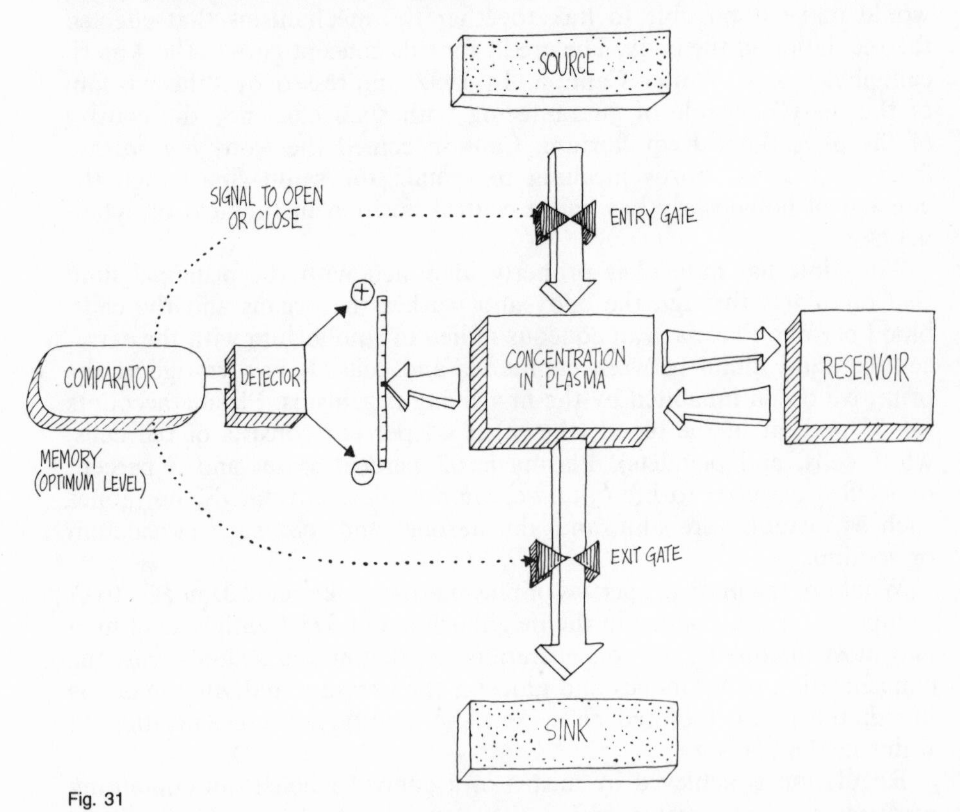

Fig. 31

When the level of calcium falls below 8.5 milligrams, a molecular detector in the tissues of the parathyroid glands sends a signal that triggers the synthesis of the parathyroid hormone. This hormone, released into the bloodstream, acts in three ways: it extracts more calcium from the bones; it slows the loss of calcium in the urine; and it increases the amount of calcium absorbed from the intestines. Consequently the level of calcium in the plasma rises. If the level should rise above 10.5 milligrams, a detector in the thyroid gland sends a signal that triggers the synthesis of the calcitonin hormone, which acts to increase the storage of calcium in the bones, thereby reducing the level of calcium in the plasma.

The regulation of other "constants" of the plasma involves the brain and behavior. One of the first detectors to be informed of internal modifications of the body is a region of the brain that plays an important role as the center of integration of the common life functions (hunger, thirst, regulation of body temperature, and sexual behavior). This center is the hypothalamus, director of instinctive functions.

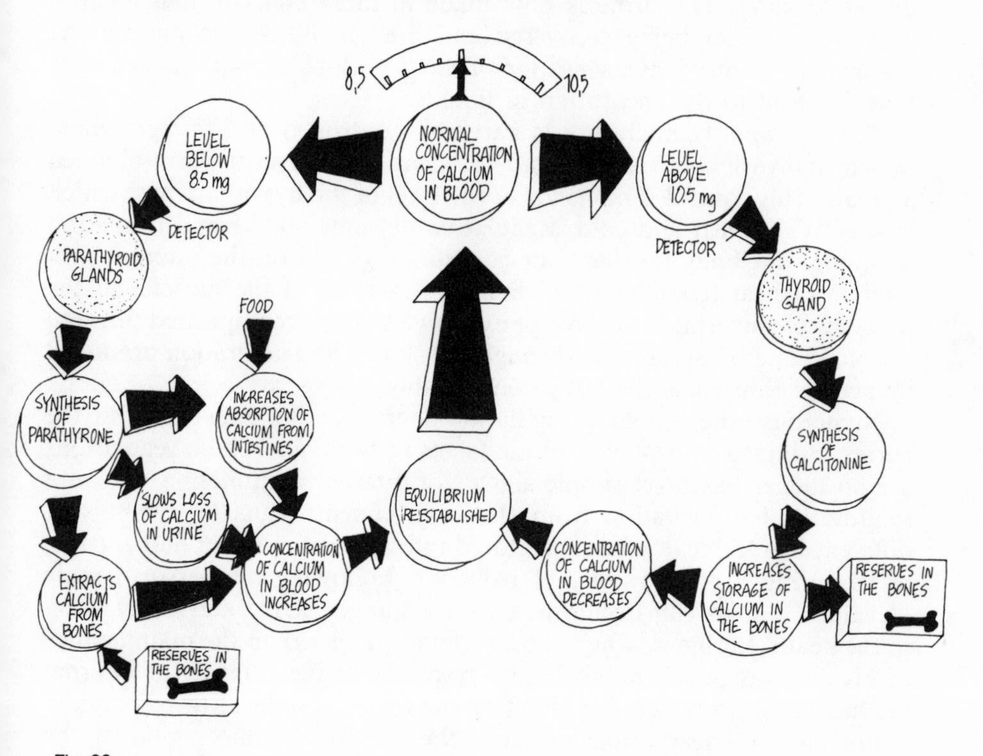

Fig. 32

Hunger. When we are hungry, the hypothalamus detects a lowering of the level of glucose, amino acids, or fatty acids in the plasma. It integrates other signals as well: body temperature, distension of the stomach. These increase the sensation of hunger. The constant of time in this regulatory mechanism is essential; a considerable period can elapse from the moment one first feels hunger to the moment of eating. A rapid response mechanism should increase the level of glucose in the blood. Thus the adrenal glands detect this disturbance of equilibrium and secrete adrenalin, which transforms glycogen reserves in the liver into readily usable glucose. In less than fifteen minutes the level of glucose begins to rise. Over a longer period (after about two hours) the secretion of hydrocortisone by the surrenal cortex permits the transformation of protein into glucose. The results of this action appear only after six to eight hours.

Thirst. When the plasma becomes too concentrated, the hypothalamus sends a signal to the hyphophyses gland, which secretes an antidiuretic hormone. This hormone releases the flow of vasopressin, which acts on the kidneys. The urine is now made in more concentrated form, a part of the water being recovered and used to dilute the plasma. At the same time, one feels a sensation of thirst—which encourages behavior that will lead to the absorption of liquids.

Temperature. Temperature is carefully controlled at 37° Centigrade in man and varies between 35° and 44° Centigrade in most warm-blooded animals. This control is managed at the level of the hypothalamus, which is sensitive to heat and cold. Regulation depends on thermal insulation (clothing, fur, body fat, heat, air conditioning) and on the internal production of heat (combustion of fats, contractions of the muscles in the presence of shivering). The loss of excessive heat is accomplished through the blood and by dissipation through the skin. The evaporation produced by perspiration cools the body considerably.

Numerous other kinds of regulation operate at the upper level of the cortex, bringing into play multiple facets of behavior. These regulations are no longer based on simple signals of internal malfunction but on a multitude of information from the environment: signs or symbols of different hierarchical value, integrated into rules of conduct and capable of triggering a great variety of behavior. Following his personal scale of values, a man can decide to go on a hunger strike and carry it on to the death. In this way he chooses a finality other than the maintenance of his own organism; he no longer responds to the "signals of internal malfunction," to which the gland or the organ is obliged to respond.

Pleasure and fear also enter into the picture. In one region of the hypothalamus there are bundles of nerve fibers that appear to play a vital role in the body's reward system. If we use electrical impulses to stimulate one of these bundles in a laboratory animal, the animal begins to eat with a relentless hunger. In the presence of an animal of the opposite sex it begins to copulate with frenzy. If we make it possible for the animal to excite and gratify itself, it will devote itself to this narcissistic activity until it is exhausted, provoking the stimuli as many as 8,000 times a day. On the other hand, any stimulation of the complementary bundles of nerve fibers induces characteristic reactions of anguish—jumping, biting, sharp cries, defensive postures.

The actual regulations of the organism require complicated circuits extending well beyond the borders of the organism and into the heart of its environment. Consider again the image of the man at work in business or industry. The quest for reward, for recognition, even for a certain gratification (in domination, in power, or simply in work well done) is combined with a constant apprehension of the discipline and the hierarchy of the company, and this has a continuous effect on the regulations of his internal equilibrium and the regulations of his equilibrium with his immediate environment. Stress, anguish, frustration, joy, pleasure, and the sense of well-being all exercise an everyday influence on hormonal regulation, on the mobilization of energy resources, and above all on our physical and mental health.

Thus the body is continuously informed of the state of its organs and its internal equilibrium, thanks to the signals that come from without and within. The brain manifests itself as the *integrator* of these signals, not as a supreme hierarchical center where decisions are made. There is no "leader" in the human body.

6. THE CELL

At the level of the living cell the concepts of the organism and society converge and illuminate each other. The metaphor of the living organism has had considerable success in its application to society; now it is the turn of the concept of society to help explain biology. "The cell, society of molecules," François Jacob writes.

At the conclusion of our opening of the Russian dolls, the last little one—that of which our knowledge is most recent—will clarify in retrospect the entire hierarchy of the levels of complexity that have led to it. The loop is going to close. From the solar energy transformed by the ecosystem to regulatory reactions that maintain the life of the cell, including the action of man on his environment, everything holds together, is connected, circles around, and overlaps.

The cell of a higher organism maintains its structure, regulates and controls its metabolic functions, grows, reproduces, performs work, exercises a specialized function within an organ, and dies. These are the functions that characterize life: self-preservation, self-regulation, self-reproduction, and the capacity to develop (Fig. 33).

Life confronts inert matter with its energizing activity. Unlike crystals, which exist and survive only in static equilibrium with the environment, the cell recurs continuously, in its most intimate composition, thanks to the flow of energy and the materials that pass through it. In spite of the molecular upheaval the cell maintains its internal organization in the face of a natural tendency to disorder. The key to this stability rests in its genetic information bank.

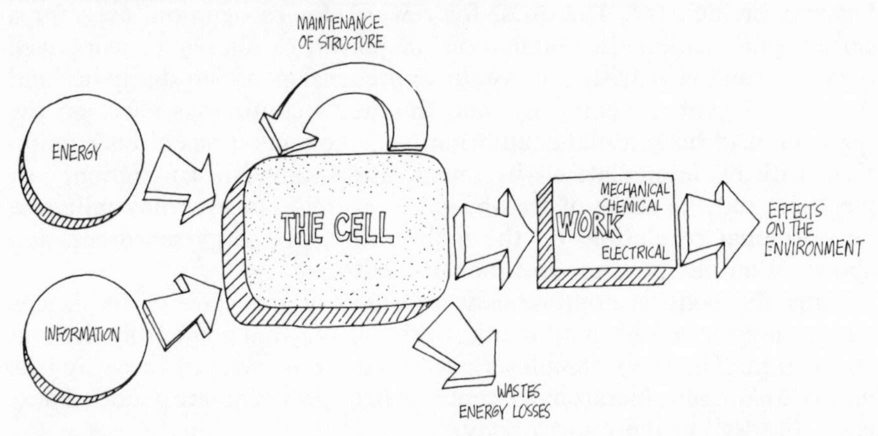

Fig. 33

Structures and functions are therefore inseparable. The maintenance of structures can be guaranteed only by the energizing activity of the functions. Structures rely on construction materials arranged according to rigorous spatial organization. The functions are exercised through the medium of a temporal organization which rests on myriads of elementary reactions that are tightly coordinated and synchronized. The cell must have *transforming agents* to maintain its structure and functions. In the cellular society these agents are molecules that form limited chemical categories.

The two main categories of chemical agents in the cell are giant molecules (macromolecules), the *proteins,* construction elements or catalysts that control cellular activity (the enzymes), and the *nucleic acids* (DNA and RNA), which store the necessary information for the assembly of proteins and enzymes and for cell reproduction.*

The other basic instruments of cellular life are signal molecules that make communication possible, energy-rich molecules, small molecules that act as building blocks, electrons and their carriers (essential in the transfer of energy), and water molecules. This entire population can be measured. In a simple cell such as a bacterium—a thousandth of a milli-

* Nucleic acids are so called because they are generally found in the nucleus of the cell.

meter long—there are from 10 to 100 billion molecules of water, 70 percent of the total composition (or population) of the cell; from 100 million to one billion molecules of average size, representing almost 500 different chemical types (sugars, fats, amino acids, pigments); and five to ten thousand distinct kinds of giant molecules of protein and enzyme that make up a population of about five million molecules. Finally, one kind of macromolecule alone contains the necessary information to direct the manufacture of all the others: deoxyribonucleic acid, or DNA.

The effectiveness of the interactions and exchanges among the various molecular instruments is assured by a small number of supramolecular organizations. Through the medium of these organizations the major functions of the cellular society are performed. The conversion of energy occurs in the *mitochondria,* the molecular power plants; the storage of energy and its reserves in the *vacuoles;* the manufacture of protein in the *ribosomes,* the assembly plants; the storage of information in the *nucleus* of the cell; and the filtration of signals to and from the outside, the protection of the cell, and the catalysis of a large number of essential reactions are performed at the level of the *membrane.*

Thus the cell appears to be a self-regulatory system, a transformer of energy, capable at all times of balancing its production in terms of its internal consumption and the energy it has at its command.

Linking the Cell and the Body

In order to relate the activity of the cell to that of the body as a whole, we must consider two complementary functions, *respiration* and *nutrition* and what happens at the cellular level.

Respiration is the basic reaction of animal life. It is a combustion in the presence of oxygen that occurs in the mitochondria. This reaction enables the cell to process food from outside sources in order to obtain the energy it needs to synthesize materials, to move about, to secrete special substances, to send electrical signals, and to reproduce. Seen from this angle, respiration appears to be a much more widespread activity than simple pulmonary ventilation, with which it is often confused.

In terms analogous to the industrial process of transformation, respiration needs fuel, combustion primer, and catalysts. The principal *fuel* of the cell is glucose; it is extracted from food by a series of converters in the digestive system and home-delivered to the cell by the distribution network of the capillaries. The *combustion primer,* obviously, is oxygen from the air, carried by the hemoglobin of the red cells and similarly home-delivered into the liquid that bathes the cells. The *catalysts* are the enzymes that speed up and control combustion and the use of the energy released. This raw energy appears first in the form of electrons.

The ultimate purpose of respiration is to recharge the "batteries" of the cell. Everything that lives uses a particular type of energy-storing molecule whose role is analogous to that of a portable battery that provides energy wherever the cell needs it to produce chemical, mechanical, or electrical activity. The molecule is ATP (adenosine triphosphate). When it has released its energy (when the battery has discharged), it is ADP (adenosine diphosphate). The cycle of combustion and extraction of electrons can be compared to a *generator* and the recharging chains of ADP and ATP to a *charger.* Figure 34 illustrates and summarizes the role of each agent.

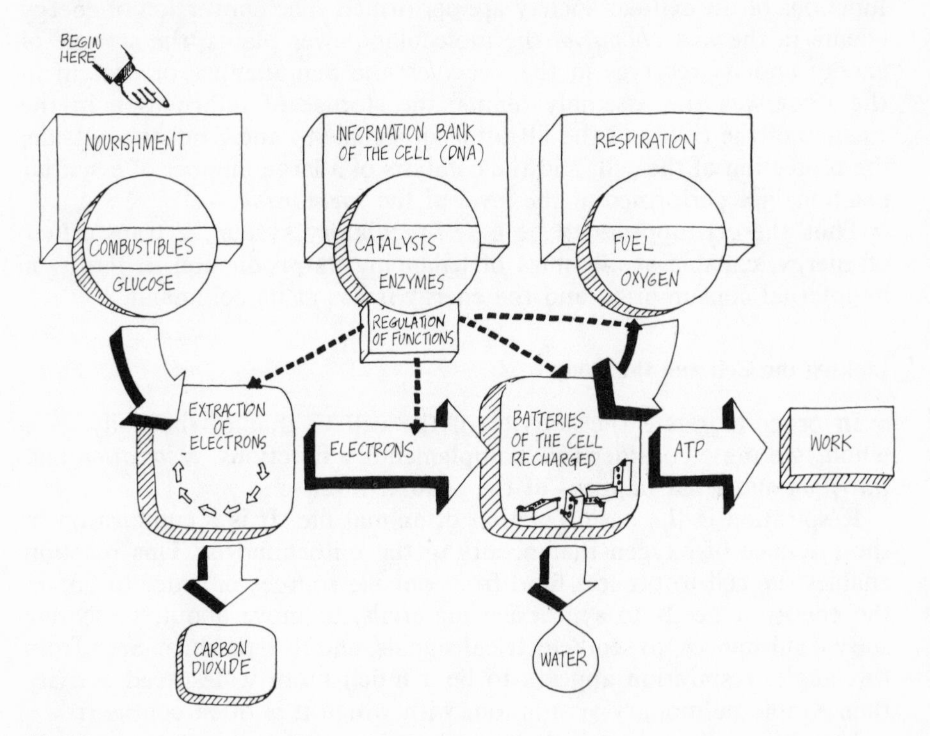

Fig. 34

This universal model will serve to explain three important aspects of the molecular functioning of the cell: the transformation and utilization of energy, the complete regulation of cellular metabolism by the enzymes, and the work of a specialized protein, hemoglobin.

The small molecules that result from digestion make up the primary raw materials of the cell. These are principally glucose, amino acids, and fatty acids. But before being used in the combustion reaction they must undergo preparation, for the generator operates only with a high-grade combustible: the molecule of activated acetic acid.

The successive reactions that insure this preparation and the extraction of electrons follow each other in strict order. At the conclusion of the series, the reactions have formed a closed cycle: the principal residue of the combustion is combined with a new molecule of activated acetic acid and reintroduced at the start of the new cycle. This cycle, which sustains the life of every complex cell, is called the Krebs cycle;* it is the "generator" of electrons.

The flow of electrons from this generator recharges the "batteries" of the cell by means of another series of reactions linked to the first.† This combined process simulates the "charger." Along the entire chain of the charger the electrons gradually lose their energy. Finally they come to the oxygen that awaits them at the end of the series (and represents the lowest level of energy in the cascade of electrons). *It is this expenditure of potential energy that powers all the machinery of life.*

What happens when violent physical effort is required of the body, as in muscular exercise or flight in the face of danger? The determining factor is the number of "run-down batteries," or the relationship between ADP and ATP (discharged and charged molecules). This relationship conditions all activity in the "generator" and in the "charger."

The mitochondrion can be compared to a service station where the batteries of a number of customers are regularly charged and the station attendant always has charged batteries on hand. Ordinarily the ratio of discharged to charged batteries is very low (about 1:100); the same is true in the cell when the ratio of ADP to ATP is very low. Then the charger is little used. The chain of electron carriers, like the generator, runs in slow motion. The demand for combustibles and oxygen is low, and the body stores glucose as glycogen and fatty acids as fats. Man sleeps, rests, and recuperates.

Then, when effort is needed, the muscles work, consuming ATP, and the "batteries" run down. The quantity of ADP (run-down batteries) increases rapidly; the ratio of ADP to ATP becomes very great (perhaps 100:1). The "service station" is flooded with calls to recharge batteries, and the recharging activity speeds up. This activity uses up more electrons and oxygen. The generator cycle turns faster and faster, consuming the stored combustibles and giving off more carbon dioxide. The quantities of glucose, amino acids, and fatty acids drop in the extracellular fluid and then in the plasma, and wastes accumulate. A full series of detectors

* The name comes from that of Sir Hans Krebs, Nobel Prize winner in medicine in 1953, who discovered the cycle.

† The chain of "electron carriers" and "oxydative phosphorylation."

in the glands and in the brain register the changes in equilibrium. Pulmonary ventilation accelerates, providing more oxygen and eliminating carbon dioxide. Heart rhythm becomes markedly faster. The blood circulates more rapidly and drains off the wastes, while the contraction of some blood vessels and the dilation of others allows an improved distribution of the blood, especially where effort is concentrated. The skin becomes red, the person becomes hot and sweats. The work of the mitochondria has had an effect throughout the *entire body* (Fig 35).

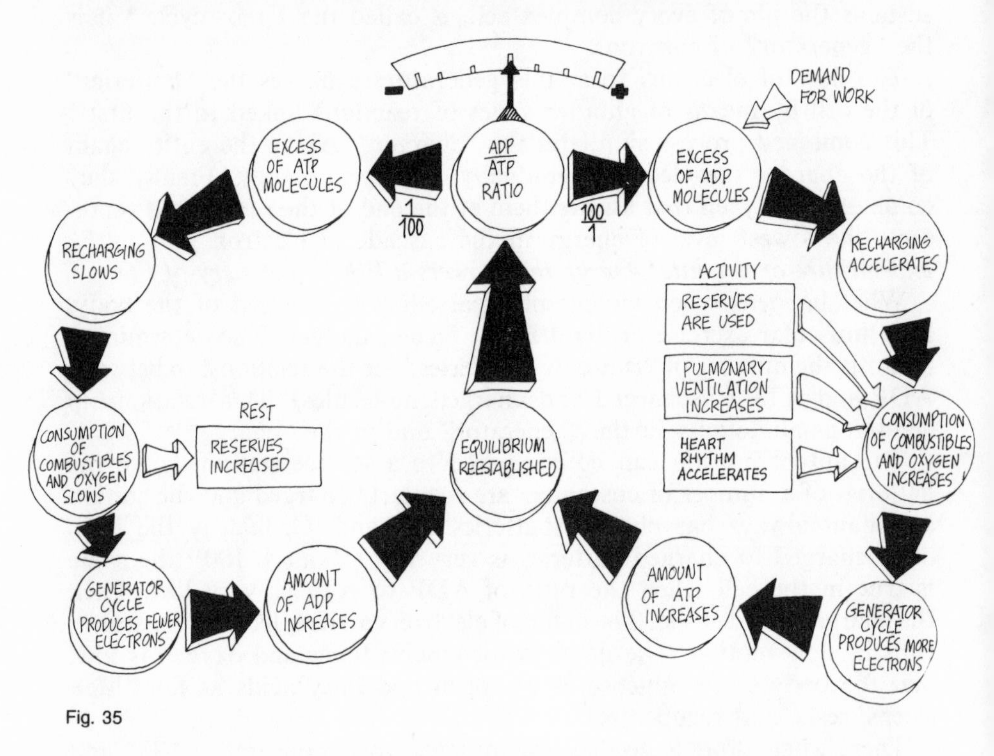

Fig. 35

After the effort is over, the drop in the level of glucose, fatty acids, and amino acids in the plasma is detected by the hypothalamus. The body becomes hungry and seeks food to recapture its strength. When activity continues for a long time—in the case of prolonged fasting, for example—glucose and reserves in the liver are not enough. The body steals from the reserves of amino acids in the proteins that build the cells. It is like burning the walls and the furniture of one's home. But one cannot risk losing more than 40 percent of one's body weight without risking death. After a certain stage the balance cannot be reestablished; the damages are irreversible.

The Work of the Enzymes

To slow or speed up a metabolic process catalyzed by a chain of enzymes (analogous to a sequence of machine tools on an assembly line), the cell exerts a simple yet harsh trick. In the short term the assembly line can be slowed or stopped at the start of each series; in the long term all or part of the assembly line can be suppressed. A speed-up in production is achieved by increasing the number of machines in each assembly line or by installing parallel lines. Thus production can be adjusted in a very short time, enabling the cell to meet a considerable demand.

The ultimate control of cellular activity must pass eventually through the production (or the blockage of production) of the enzymes. This production takes place in the assembly shops of the cell, but the original plans for all the special features of enzymes that the cell needs *never leave the nucleus of the cell.* Thus if the plans cannot be copied inside the nucleus, no enzyme will be assembled in the workshops. As long as the "copying machine" functions, production continues without hindrance.

Among the genes in the nucleus of the cell there is a battery of switches, the repressor molecules, that can control the operation of the machine that copies the plans. Each repressor recognizes a specific signal that orders it to stop or to carry on the copying of the plans of enzymes specialized in a given task. This regulating signal is generally a small molecule that attaches itself to the appropriate repressor and activates or deactivates it (Fig. 36).

Here we see in action the important signal molecules on which relies a large part of the information that circulates in the cells and throughout the body. These molecules are recognized by specialized detectors located, as we have seen, in the glands and organs. They turn on or off chemical switches (like the repressors) and they trigger or block the synthesis of enzymes and (indirectly) hormones and other molecules essential to life.

The functioning of the repressors and the functioning of the enzymes are possible only because of recognition mechanisms that act between nucleic acids, proteins, and the regulatory molecules. The recognition of information according to the shape of the molecules (their morphology) is very general; it is the basis of the universal language of internal communication used by all cells.

Hemoglobin is an extraordinary machine, a veritable "molecular lung." Its purpose is to carry oxygen from the lungs to the tissues, via the arteries and the capillary network. It uses the network of veins to return directly or indirectly to the lungs the carbon dioxide remaining in the tissues. On one side the blood is bright red; on the other, dark brown.

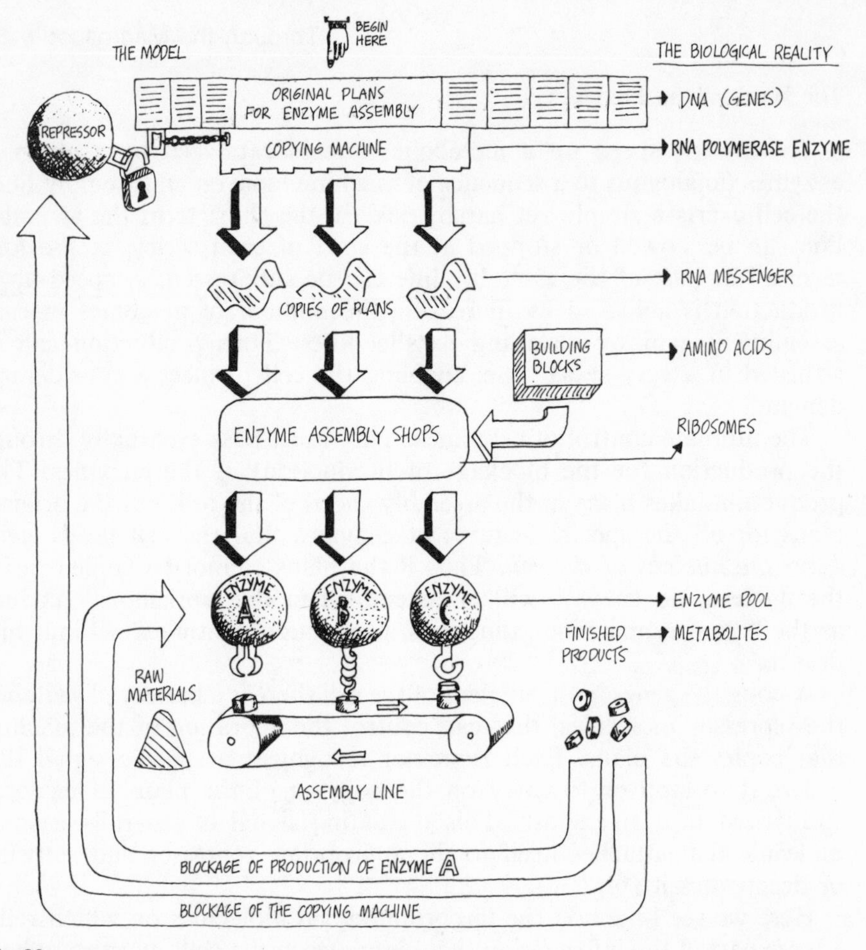

Fig. 36

The hemoglobin must give up its oxygen at the right place in the tissue and not return it to the lungs. And this is one of the paradoxical qualities of this molecule: it is capable of taking on oxygen as easily as it is of discharging it.

As in the case of most proteins, the properties of hemoglobin depend on its molecular structure and its "anatomy." It is made up of four blocks, each linked with the other in a compact structure by a sort of molecular staple. Each block is a protein, globin, composed of a chain of amino acids all linked to one another. Near the center of each block lies a molecule, flat as a disk, which contains an atom of iron in its center. This molecule is a pigment, the heme, that gives blood its red color (Fig. 37).

This pigment and its atom of iron constitute an "active site" that is able to recognize and to capture oxygen molecules. There are four such sites, so the hemoglobin can bind four molecules of oxygen. As the absorption of oxygen in the pulmonary tissue proceeds, molecular staples pop out. Thus the four blocks modify their arrangement in space, which makes the absorption of other oxygen molecules much easier.

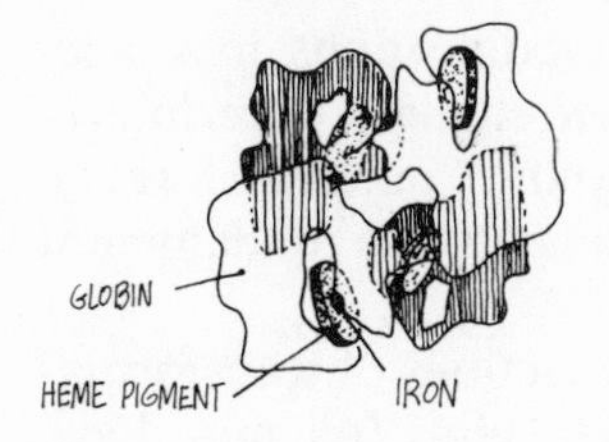

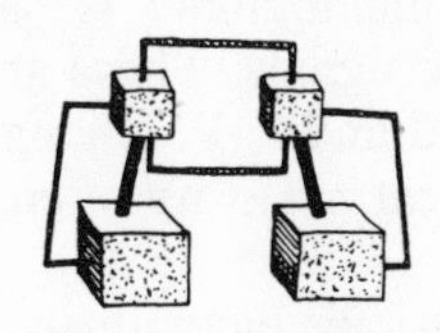

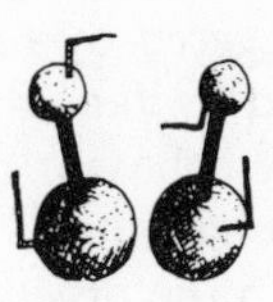

Fig. 37

The entire working of hemoglobin rests on a simple property of iron: in the presence of oxygen its diameter decreases by about 13 percent. This diminution in size allows it to lodge more readily in the plane of the flat pigment molecule. The light movement that follows it is amplified by the chain to which the iron is attached, which serves as a series of levers and springs. Tension can make one of the clips pop—a little like a snap fastener that pops from its place; consequently a block changes slightly in shape and position in relation to others. This makes it easier for the next block to bind another molecule of oxygen, and the process continues through the succeeding blocks.

Hemoglobin discharges all its oxygen in the cellular tissues, the more readily as the molecules of oxygen are freed. Because of this mechanism, hemoglobin pumps oxygen in only one direction, from the lungs to the tissues. In fact in every organism there is a balance between two kinds of hemoglobin, the deoxidized form and the oxidized form (Fig. 38).

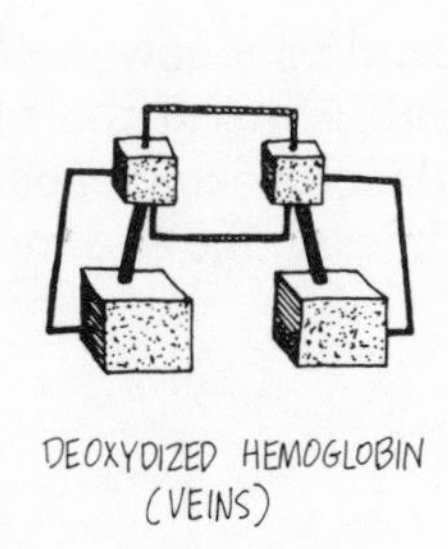

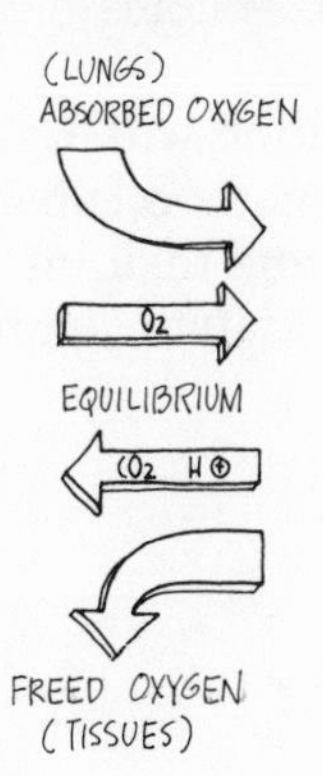

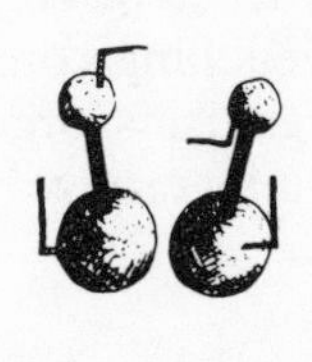

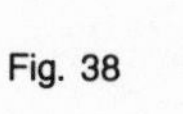

Fig. 38

Everything that stabilizes the deoxidized form allows more oxygen to be discharged by displacing the equilibrium in that direction. This is the role played by regulatory signals that are present in cell tissues—like the molecule of carbon dioxide or the acid ions (H^+), the principal wastes from the activity of the cell (see p. 51). Now we can understand why every effort makes us breathe more rapidly.

The activity of the hemoglobin is based on modifications in shape, triggered by regulatory signals. These are *allosteric* changes (a term created by Jacques Monod and J.-P. Changeux, meaning "different form"). The behavior of the great majority of enzymes rests on this fundamental mechanism.

In the preceding pages we have illustrated the reactions of transformation of energy and of regulation that characterize life. You may have noticed that in the diagram on page 50 some arrows arrive from nowhere and lead nowhere (glucose and carbon dioxide at left, oxygen and water at right). In fact what is lacking in this chain of life is an essential link, the green vegetable cell. This cell manufactures energy-rich glucose during periods of photosynthesis with solar energy, water, and the carbon dioxide released by animals. This is what introduces into the atmosphere the oxygen necessary for respiration. Thus the loop closes on the ecosystem and solar energy.

From the main cycles of life to the tiniest molecular cogs and the subtle play of electrons, this last plunge into the heart of the cell has clarified, I hope, the unity of the fundamental mechanisms of nature and society. The enzyme, the cell, and the organ represent—each at its own level—the catalysts of the many functions that maintain, regulate, or transform the organization on which their lives depend. Man, too, is one of these catalysts. To understand better how they act within their own organisms can lead man to better behavior, from within, in the transformation of the complex systems on which he depends: business, the city, society.

In order to be effective his action will have to depend on a new method of approaching complexity, a method capable of embracing at the same time organisms, organizations, and their interdependencies—and capable, too, of integrating, beyond the analytic approach, "knowledge and meaning."

Two

The Systemic Revolution: A New Culture

1. HISTORY OF A GLOBAL APPROACH

The fundamental concepts that recur most often in the biological, ecological, and economic models of the preceding chapter can easily be grouped into several major categories: energy and its use; flows, cycles, and stocks; communication networks; catalysts and transforming agents; the readjustment of equilibriums; stability, growth, and evolution. And, above all, the concept of the system—living system, economic system, ecosystem—that binds together all the others.

Each of these concepts applies to the cell as it does to the economy, to an industrial company as it does to ecology. Beyond the vocabulary, the analogies, and the metaphors there appears to exist a common approach that makes it possible to understand better and describe better the organized complexity.

The Systemic Approach

This unifying approach does indeed exist. It was born in the course of the last thirty years from the cross-fertilization of several disciplines—biology, information theory, cybernetics, and systems theory. It is not a new concept; what is new is the integration of the disciplines that has come about around it. This transdisciplinary approach is called the *systemic approach,* and this is the approach that I present here in the concept of the macroscope. It is not to be considered a "science," a "theory," or a "discipline," but a *new methodology that makes possible the collection and organization of accumulated knowledge in order to increase the efficiency of our actions.*

The systemic approach, as opposed to the analytical approach, includes the totality of the elements in the system under study, as well as their interaction and interdependence.

The systemic approach rests on the conception of system. While often vague and ambiguous, this conception is nevertheless being used today in an increased number of disciplines because of its ability to unify and integrate.

According to the most widely used definition, "a system is a set of interacting elements that form an integrated whole." A city, a cell, and a body, then, are systems. And so are an automobile, a computer, and a washing machine! Such a definition can be too general. Yet no *definition* of the word *system* can be entirely satisfying; it is the *conception* of system that is fertile—if one measures its extent and its limits.

The limits are well known. Applied too easily, the systems concept is often used wildly in the most diverse areas: education, management, data processing, politics. For numerous specialists it is only an empty notion: trying to say everything, it evokes nothing in the end.

Yet its reach cannot be held to the precision of definitions; the concept of system is not readily confined. It reveals and enriches itself only in the indirect illumination of the many clusters of analogous, modeled, and metaphoric expression. The concept of system is the crossroads of the metaphors; ideas from all the disciplines travel there. Reaching beyond single analogies, this circulation makes possible the discovery of what is common among the most varied systems. It is no longer a matter of reducing one system to another, better-known one (economics to biology, for example); nor does it mean transposing knowledge from a lower level of complexity to another level. It is a question of identifying *nonvariants*—that is, the general, structural, and functional principles—and being able to apply them to one system as well as another. With these principles it becomes possible to organize knowledge in models that are easily transferred and then to use some of these models in thought and action. Thus the concept of system appears in two complementary aspects: it enables the organization of knowledge and it renders action more efficient.

In concluding this introduction to the concept of system, we need to locate the systemic approach with respect to other approaches with which it is often confused.

- The systemic approach embraces and goes beyond the *cybernetics* approach (N. Wiener, 1948), whose main objective is the study of control in living organisms and machines.
- It must be distinguished from *General Systems Theory* (L. von Bertalanffy, 1954), whose purpose is to describe in mathematical language the totality of systems found in nature.
- It turns away from *systems analysis,* a method that represents only one tool of the systemic approach. Taken alone, it leads to the reduction of a system to its components and its elementary interactions.
- The systemic approach has nothing to do with *a systematic approach* that

confronts a problem or sets up a series of actions in sequential manner, in a detailed way, forgetting no element and leaving nothing to chance.

Perhaps one of the best ways of seeing the strength and the impact of the systemic approach is to follow its birth and development in the lives of men and institutions.

The Search for New Tools

The process of thought is at once analytic and synthetic, detailed and holistic. It rests on the reality of facts and the perfection of detail. At the same time, it seeks factors of integration, catalytic elements for invention and imagination. At the very moment that man discovered the simplest elements of matter and life, he tried, with the help of the famous metaphors of the "clock," the "machine," the "living organism," to understand better the interactions between these elements.

Despite the strengths of these analogical models, thought is dispersed in a maze of disciplines each secluded one from another by communication-tight enclosures. The only way to master these numbers, to understand and predict the behavior of the multitudes made up of atoms, molecules, or individuals, is to reduce them to statistics and to derive from them the laws of unorganized complexity.

The theory of probability, the kinetic theory of gases, thermodynamics, and population statistics all rely on unreal, "ghostly" phenomena, on useful but ideal simplifications that are almost never found in nature. Theirs is the universe of the homogeneous, the isotope, the additive, and the linear; it is the world of "perfect" gases, of "reversible" reactions, of "perfect" competition.

In biology and in sociology, phenomena integrate duration and irreversibility. Interactions between elements count as much as the elements themselves. Thus we need new tools with which to approach organized complexity, interdependence, and regulation.

The tools emerged in the United States in the 1940s from the cross-fertilization of ideas that is common in the melting pot of the large universities.

In illustrating a new current of thought, it is often useful to follow a thread. Our thread will be the Massachusetts Institute of Technology (MIT). In three steps, each of about ten years, MIT was to go from the birth of cybernetics to the most critical issue, the debate on limits to growth. Each of these advances was marked by many travels back and forth—typical of the systemic approach—between machine, man, and society. In the course of this circulation of ideas there occurred transfers of method and terminology that later fertilized unexplored territory.

In the forties the first step forward led from the machine to the living organism, transferring from one to the other the ideas of feedback and finality and opening the way for automation and computers (Fig. 39).

Fig. 39

In the fifties it was the return from the living organism to the machine, with the emergence of the important concepts of memory and pattern recognition, of adaptive phenomena and learning, and new advances in bionics: artificial intelligence and industrial robots.* There was also a return from the machine to the living organism, which accelerated progress in neurology, perception, the mechanisms of vision (Fig. 40).

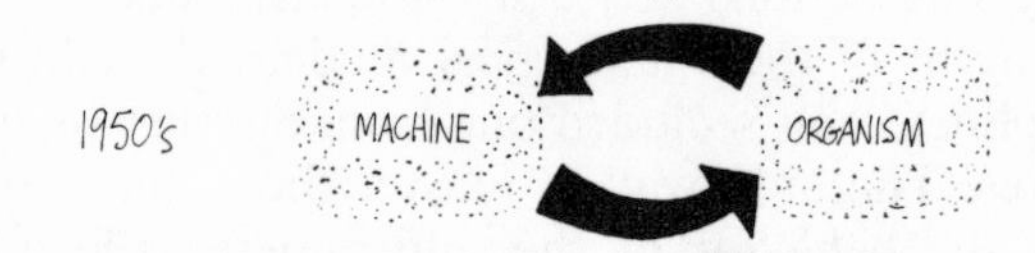

Fig. 40

In the sixties MIT saw the extension of cybernetics and system theory to industry, society, and ecology (Fig. 41).

Fig. 41

Three men can be regarded as the pioneers of these great breakthroughs: the mathematician Norbert Wiener, who died in 1964; the neurophysiologist Warren McCulloch, who died in 1969; and Jay Forrester, professor at the Sloan School of Management at MIT.

There are of course other men, other research teams, other universities—in the United States as well as in the rest of the world—that have contributed to the advance of cybernetics and system theory. I will mention them whenever their course of research blends with that of the MIT teams.

---

* Bionics attempts to build electronic machines that imitate the functions of certain organs of living beings.

"Intelligent" Machines

Norbert Wiener had been teaching mathematics at MIT since 1919. Soon after his arrival there he had become acquainted with the neurophysiologist Arturo Rosenblueth, onetime collaborater of Walter B. Cannon (who gave homeostasis its name; see p. 43) and now at Harvard Medical School. Out of this new friendship would be born, twenty years later, cybernetics. With Wiener's help Rosenblueth set up small interdisciplinary teams to explore the no man's land between the established sciences.

In 1940 Wiener worked with a young engineer, Julian H. Bigelow, to develop automatic range finders for antiaircraft guns. Such servomechanisms are able to predict the trajectory of an airplane by taking into account the elements of past trajectories. During the course of their work Wiener and Bigelow were struck by two astonishing facts: the seemingly "intelligent" behavior of these machines and the "diseases" that could affect them. Theirs appeared to be "intelligent" behavior because they dealt with "experience" (the recording of past events) and predictions of the future. There was also a strange defect in performance: if one tried to reduce the friction, the system entered into a series of uncontrollable oscillations.

Impressed by this disease of the machine, Wiener asked Rosenblueth whether such behavior was found in man. The response was affirmative: in the event of certain injuries to the cerebellum, the patient cannot lift a glass of water to his mouth; the movements are amplified until the contents of the glass spill on the ground. From this Wiener inferred that in order to control a finalized action (an action with a purpose) the circulation of information needed for control must form "a closed loop allowing the evaluation of the effects of one's actions and the adaptation of future conduct based on past performances." This is typical of the guidance system of the antiaircraft gun, and it is equally characteristic

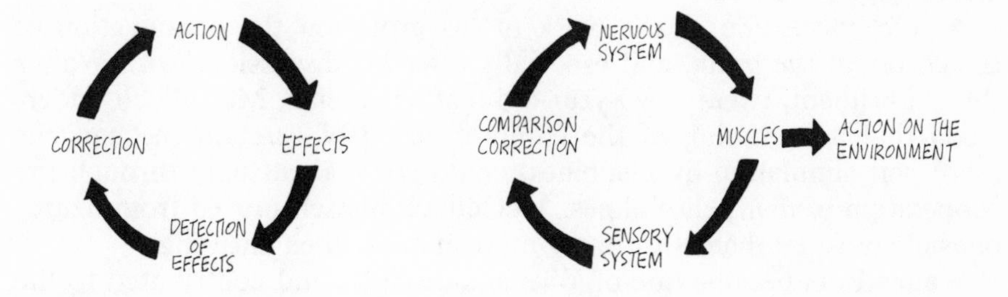

Fig. 42

of the nervous system when it orders the muscles to make a movement whose effects are then detected by the senses and fed back to the brain (Fig. 42).

Thus Wiener and Bigelow discovered the closed loop of information necessary to correct any action—the negative feedback loop—and they generalized this discovery in terms of the human organism.

During this period the multidisciplinary teams of Rosenblueth were being formed and organized. Their purpose was to approach the study of living organisms from the viewpoint of a servomechanisms engineer and, conversely, to consider servomechanisms with the experience of the physiologist. An early seminar at the Institute for Advanced Study at Princeton in 1942 brought together mathematicians, physiologists, and mechanical and electrical engineers. In light of its success, a series of ten seminars was arranged by the Josiah Macy Foundation. One man working with Rosenblueth in getting these seminars under way was the neurophysiologist Warren McCulloch, who was to play a considerable role in the new field of cybernetics. In 1948 two basic publications marked an epoch already fertile with new ideas: Norbert Wiener's *Cybernetics, or Control and Communication in the Animal and the Machine;* and *The Mathematical Theory of Communication* by Claude Shannon and Warren Weaver. The latter work founded information theory.

The ideas of Wiener, Bigelow, and Rosenblueth caught fire like a trail of powder. Other groups were formed in the United States and around the world, notably the Society for General Systems Research, whose publications deal with disciplines far removed from engineering, such as sociology, political science, and psychiatry.

The seminars of the Josiah Macy Foundation continued, opening to new disciplines: anthropology with Margaret Mead, economics with Oskar Morgenstern. Mead urged Wiener to extend his ideas to society as a whole. Above all, the period was marked by the profound influence of Warren McCulloch, director of the Neuropsychiatric Institute at the University of Illinois.

At the conclusion of the work of his group on the organization of the cortex of the brain, and especially after his discussions with Walter Pitts, a brilliant, twenty-two-year-old mathematician, McCulloch understood that a beginning of the comprehension of cerebral mechanisms (and their simulation by machines) could come about only through the cooperation of many disciplines. McCulloch himself moved from neurophysiology to mathematics, from mathematics to engineering.

Walter Pitts became one of Wiener's disciples and contributed to the exchange of ideas between Wiener and McCulloch; it was he who succeeded in convincing McCulloch to install himself at MIT in 1952 with his entire team of physiologists.

From Cybernetics to System Dynamics

In this famous melting pot, ideas boiled. From one research group to another the vocabularies of engineering and physiology were used interchangeably. Little by little the basics of a common language of cybernetics was created: learning, regulation, adaptation, self-organization, perception, memory. Influenced by the ideas of Bigelow, McCulloch developed an artificial retina in collaboration with Louis Sutro of the laboratory of instrumentation at MIT. The theoretical basis was provided by his research on the eye of the frog, performed in 1959 in collaboration with Lettvin, Maturana, and Pitts. The need to make machines imitate certain functions typical of living organisms contributed to the speeding up of progress in the understanding of cerebral mechanisms. This was the beginning of bionics and the research on artificial intelligence and robots.

Paralleling the work of the teams of Wiener and McCulloch at MIT, another group tried to utilize cybernetics on a wider scope. This was the Society for General Systems Research, created in 1954 and led by the biologist Ludwig von Bertalanffy. Many researchers were to join him: the mathematician A. Rapoport, the biologist W. Ross Ashby, the biophysicist N. Rashevsky, the economist K. Boulding. In 1954 the *General Systems Yearbooks* began to appear; their influence was to be profound on all those who sought to expand the cybernetic approach to social systems and the industrial firm in particular.

During the fifties a tool was developed and perfected that would permit organized complexity to be approached from a totally new angle—the computer. The first ones were ENIAC (1946) and EDVAC or EDSAC (1947). One of the fastest was Wirlwind II, constructed at MIT in 1951. It used—for the first time—a superfast magnetic memory invented by a young electronics engineer from the servomechanisms laboratory, Jay W. Forrester.*

As head of the Lincoln Laboratory, Forrester was assigned by the Air Force in 1952 to coordinate the implementation of an alert and defense system, the SAGE system, using radar and computers for the first time.† Its mission was to detect and prevent possible attack on American territory by enemy rockets. Forrester realized the importance of the systemic approach in the conception and control of complex organizations involving men and machines in "real time": the machines had to be capable of making vital decisions *as the information arrived.*

In 1961, having become a professor at the Sloan School of Management

---

* IBM subsequently used such memories in all its computers. This type of memory (for which Forrester still holds all major patents) is in the process of being replaced by semiconductor memories. (The former type is still found in most computers today.)

† SAGE: Semi-Automatic Ground Equipment.

at MIT, Forrester created Industrial Dynamics. His object was to regard all industries as cybernetics systems in order to simulate and to try to predict their behavior.

In 1964, confronted with the problems of the growth and decay of cities, he extended the industrial dynamics concept to urban systems (Urban Dynamics). Finally, in 1971, he generalized his earlier works by creating a new discipline, system dynamics, and published *World Dynamics.* This book was the basis of the work of Dennis H. Meadows and his team on the limits to growth. Financed by the Club of Rome, these works were to have worldwide impact under the name *MIT Report.*

Figure 43 brings together the researchers and teams mentioned in the preceding pages and recalls the main lines of thought opened up by their work.

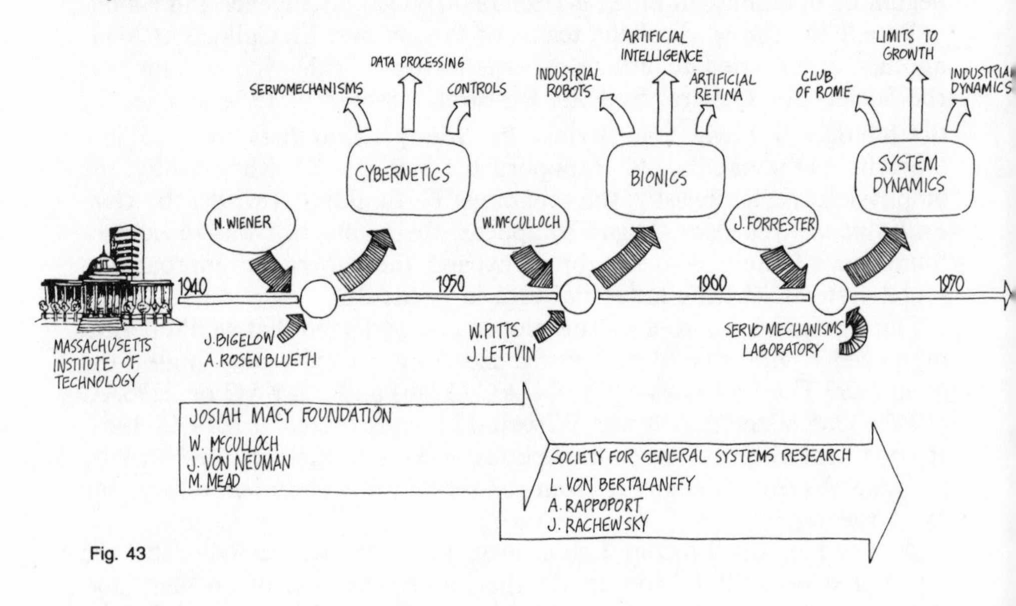

Fig. 43

2. WHAT IS A SYSTEM?

The systemic approach depends on cybernetics and system theory. Perhaps it will be useful here to recall a few definitions. Cybernetics is the discipline that studies communication and control in living beings and in the machines built by man. A more philosophical definition, suggested by Louis Couffignal in 1958, considers cybernetics as *"the art of assuring efficiency of action."* The word *cybernetics* was reinvented by Norbert Wiener in 1948 from the Greek *kubernetes,* pilot, or rudder.*

* The word was first used by Plato in the sense of "the art of steering" or "the art of government." In 1834 Ampère used the word *cybernetics* to denote "the study of ways of governing."

One of the very first cybernetics mechanisms to control the speed of the steam engine, invented by James Watt and Matthew Boulton in 1788, was called a *governor,* or a ball regulator. Cybernetics has in fact the same root as government: the art of managing and directing highly complex systems.

There are definitions of the word *system* other than that given at the beginning of this chapter. This is the most complete: *"a system is a set of elements in dynamic interaction, organized for a goal."*

The introduction of finality (the goal of the system) in this definition may be surprising. We understand that the purpose of a machine has been defined and specified by man; but how does one speak of the purpose of a system like the cell? There is nothing mysterious about the "goal" of the cell. It suggests no scheme; it declares itself *a posteriori:* to maintain its structure and replicate itself. The same applies to the ecosystem. Its purpose is to maintain its equilibrium and permit the development of life. No one has set the level of the concentration of oxygen in the air, the average temperature of the earth, the composition of the oceans. They are maintained, however, within very strict limits.

The preceding definition is distinct from that of a certain structuralist tendency, for which a system is a closed structure. Such a structure cannot evolve but passes through phases of collapse due to an internal disequilibrium.

In fact such definitions, as we said, are too general to be truly useful. They do not allow clarification of such ambiguities of expression as "a political system," "a computer system," and "a system of transportation." On the other hand, it seems to be much more profitable to enrich the concept of systems by describing in the most general way the principal characteristics and properties of systems, no matter what level of complexity they may belong to.*

Open Systems and Complexity

Each of the Russian dolls described in the first chapter is an open system of high complexity. These are important concepts that we must examine.

An *open system* is in permanent relation with its environment, or, in general terms, with its ecosystem. It exchanges energy, matter, and information used in the maintenance of its organization to counter the ravages of time. It dumps into the environment entropy, or "used" energy. By virtue of the flow of energy through the system—and despite the accumulation of entropy in the environment—the entropy in an open

---

* I do not consider here systems of concept or mechanical systems run by man, but instead systems of high complexity, such as living, social, or ecological systems.

system is maintained at a relatively low level. This is another way of saying that the organization of the system is maintained. Open systems can decrease entropy locally and can even evolve toward states of higher complexity.

An open system, then, is a sort of reservoir that fills and empties at the same speed: water is maintained at the same level as long as the volume of water entering and leaving remain the same (Fig. 44).

Fig. 44

To emphasize the generality and importance of the concept of the open system, I have used the same kind of basic diagram for the industrial firm, the city, the living organism, and the cell. One must keep in mind that open system and ecosystem (or environment) are in constant interaction, *each one modifying the other and being modified in return* (Fig. 45).

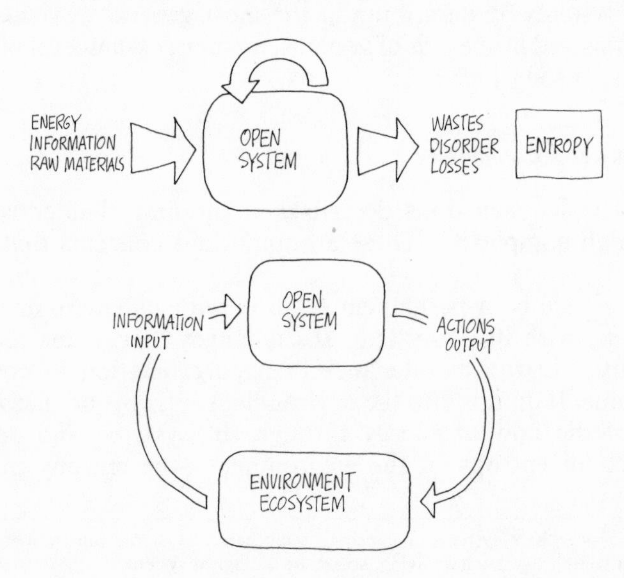

Fig. 45

A *closed system* exchanges neither energy nor matter nor information with its environment; it is totally cut off from the outside world. The system uses its own internal reserve of potential energy. As its reactions take place, entropy advances irreversibly. When the thermodynamic equilibrium is reached, entropy is maximized: the system can no longer produce work. Classical thermodynamics considers only closed systems, but a closed system is an abstraction of physicists, a simplification that has made possible the fundamental laws of physical chemistry (Fig. 46).

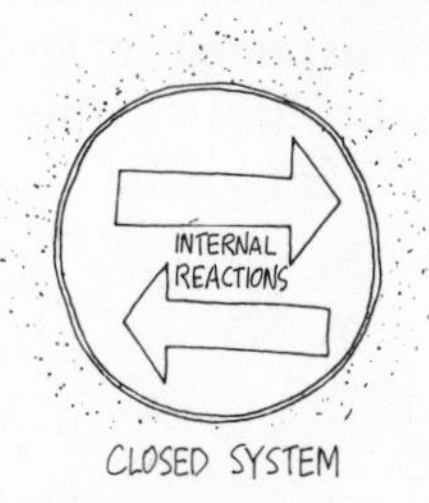

Fig. 46

How to define *complexity?* Or, to avoid definitions, how to illustrate and enrich the significance of the concept? Two factors are important: the variety of elements and the interaction between elements.

A gas, a simple system, is made up of similar elements (molecules of oxygen, for example) that are unorganized and display weak interactions. On the other hand, a cell—a complex system—includes a large variety of organized elements in tight interaction with one another. One could illustrate the concept of complexity with several points.

- A complex system is made up of a large *variety* of components or elements that possess specialized functions.
- These elements are organized in internal hierarchical *levels* (in the human body, for example, cells, organs, and systems of organs).
- The different levels and individual elements are linked by a great variety of *bonds.* Consequently there is a high concentration of interconnections.*
- The interactions between the elements of a complex system are of a particular type; they are *nonlinear* interactions.

The effects of simple linear interactions can be described by mathematical relationships in which the variables are increased or diminished by a *constant* quantity (as in the case of a car moving at the same average speed on a highway) (Fig. 47).

* The variety defined by W. Ross Ashby is "the number of different elements that make up a system or the number of different relationships between these elements or the number of different states of these relationships." The variety of a relatively simple system, made up of seven elements connected by two-way relationships and having two different sets of conditions, will be expressed by the enormous number of 2^{42}. What can be said of these interactions woven together in the heart of the cellular population (see p. 48) and in much greater number in the heart of society?

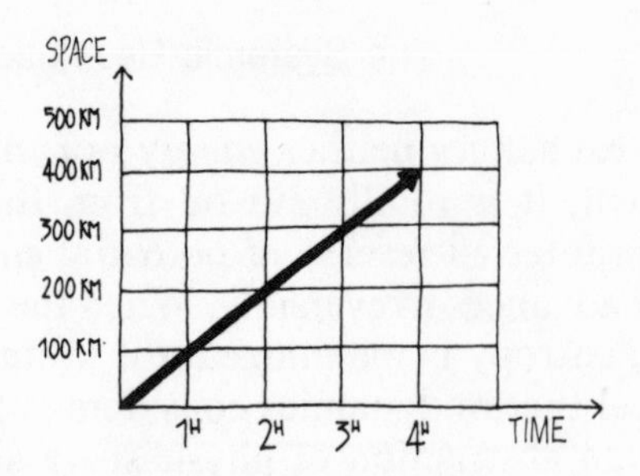

Fig. 47

However, in the case of nonlinear interactions the variables are multiplied or divided by coefficients which are themselves functions of other variables. This is the case of exponential growth (the quantity plotted on the vertical axis doubles by unit of time) or of an S curve (rapid growth following stabilization) (Fig. 48).

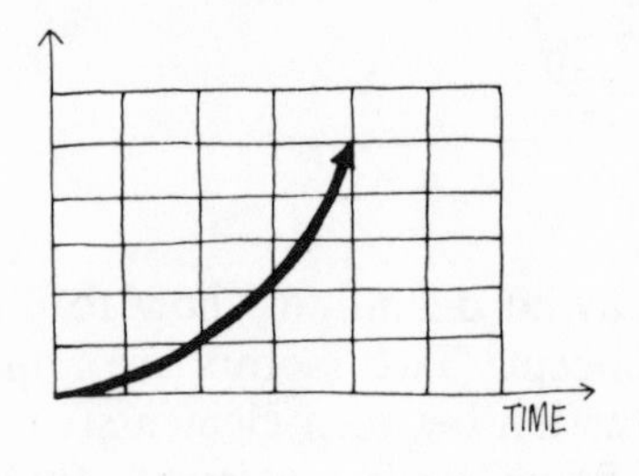

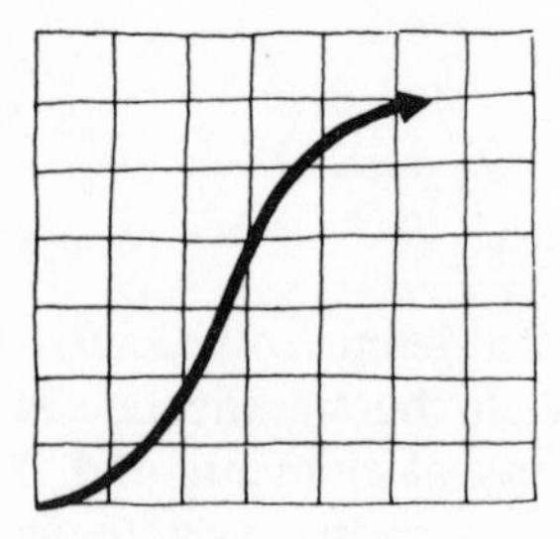

Fig. 48

Another example of a nonlinear relationship is the response of enzymes to different concentrations of substrate (molecules that they transform). In some cases, in the presence of inhibitors, the speed of transformation is slow. In others, in the presence of activators, the reaction is rapid up to the saturation of the active sites. In Figure 49 below this situation is expressed in curves that show the number of molecules transformed (1) in the presence of an inhibitor, (2) in the presence of an activator, and (3) according to the relative concentration of inhibitors and activators.

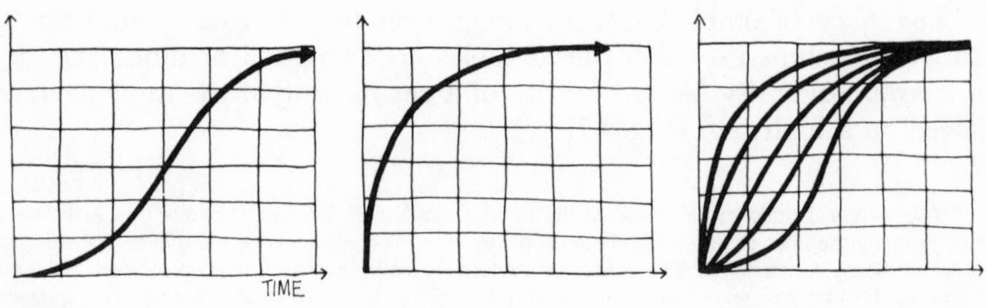

Fig. 49

Linked to the concept of complexity are those of the variety of elements and interactions, of the nonlinear aspect of interactions, and of organized totality. There follows a very special behavior of complex systems that is difficult to predict. It is characterized by the emergence of new properties and great resistance to change.

Structural and Functional Aspects of Systems

Two groups of characteristic features make it possible to describe in a very general way the systems that can be observed in nature. The first group relates to their structural aspect, the second to their functional aspect.

The structural aspect concerns the organization in space of the components or elements of a system; this is spatial organization. The functional aspect concerns process, or the phenomena dependent on time (exchange, transfer, flow, growth, evolution); this is temporal organization.

It is easy to connect the structural and functional elements by using a simple graphic illustration, a "symbolic meccano," which makes it possible to construct models of different systems and to understand better the role of interactions.*

The principal structural characteristics of every system are:

A *limit* that describes the boundaries of the system and separates it from the outside world. It is the membrane of a cell, the skin of a body, the walls of a city, the borders of a country.

Elements or components that can be counted and assembled in categories, families, or populations. They are the molecules of a cell, the inhabitants of a city, the personnel of an industrial firm and its machines, institutions, money, goods.

Reservoirs in which the elements can be gathered and in which energy, information, and materials are stored. In the first chapter numerous examples were given: reservoirs in the atmosphere and in the sediments; reservoirs of hydrocarbons; stores of capital and technology; memory banks, libraries, films, tape recordings; the fats of the body, glycogen of the liver. The symbolic representation of a reservoir is a simple rectangle.

A communication network that permits the exchange of energy, matter, and information among the elements of the system and between different reservoirs. This network can assume the most diverse forms: pipes, wires, cables, nerves, veins, arteries, roads, canals, pipelines, electric transmission lines. The network is represented in diagrams by lines and dotted lines that link the reservoirs or other variables of the model.

---

* This symbolic representation was inspired by the one developed by Jay Forrester and his group at MIT in simulation models.

The principal functional characteristics of every system are:

Flows of energy, information, or elements that circulate between reservoirs. They are always expressed in quantities over periods of time. There are flows of money (salaries in dollars per month), finished products (number of cars coming off the assembly line by the day or the month), people (number of travelers per hour), information (so many bits of information per microsecond in a computer). Flows of energy and materials raise or lower the levels in the reservoirs. They circulate through the networks of communication and are represented symbolically by a heavy black arrow (flows of information are indicated by a dotted-line arrow). Information serves as a basis for making the decisions that move the flows which maintain reserves or raise and lower the levels of the reservoirs.

Valves that control the volume of various flows. Each valve is a center of decision that receives information and transforms it into action: a manager of industry, an institution, a transforming agent, a catalyst such as an enzyme. Valves can increase or diminish the intensity of flows. Their symbolic representation is that of a valve or a faucet superimposed on a line of flow (Fig. 50).

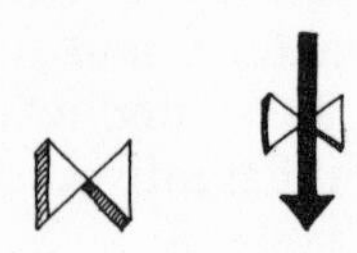

Fig. 50

Delays that result from variations in the speed of circulation of the flows, in the time of storage in the reservoirs, or in the "friction" between elements of the system. Delays have an important role in the phenomena of amplification or inhibition that are typical of the behavior of complex systems.

Feedback loops or information loops that play a decisive part in the behavior of a system through integrating the effects of reservoirs, delays, valves, and flows. Numerous examples of feedback were given in the first chapter: population control, price equilibriums, the level of calcium in the plasma (see pp. 10, 25, 43). There are two kinds of feedback loops. Positive feedback loops contain the dynamics for change in a system (growth and evolution, for example); negative feedback loops represent control and stability, the reestablishment of equilibriums and self-maintenance.

The model in Figure 51 combines all the structural and fundamental symbols described above. And here it is possible to illustrate the difference between a positive and a negative feedback loop. If the information received at the level of the reservoir indicates that the level is rising, the decision to open the valves wider will allow overflow; if the level is falling, the decision to reduce the outflow will lead to a rapid drying up of the reservoir. This is the work of a positive feedback loop, working toward infinity or toward zero. In contrast, the decision to diminish the flow when the level increases (and the inverse) maintains the level at a constant depth. This is the work of a negative feedback loop.

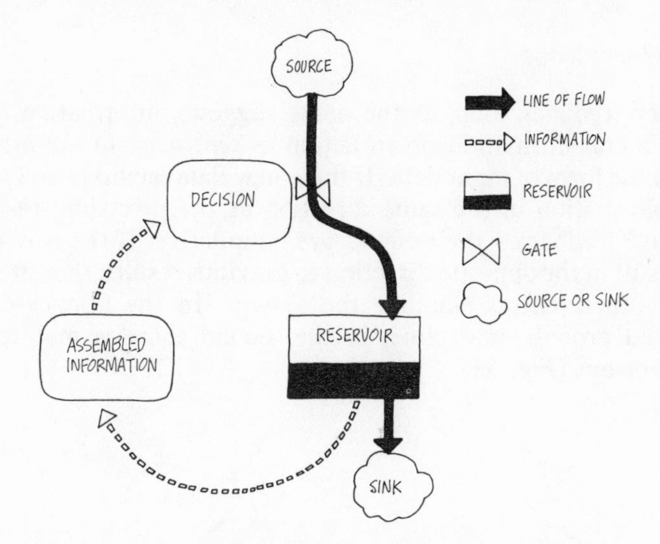

Fig. 51

3. SYSTEM DYNAMICS: THE INTERNAL CAUSES

The basic functioning of systems depends on the interplay of feedback loops, flows, and reservoirs. They are three of the most general concepts of the systemic approach, and they are the keys to the juxtaposition of very different areas from biology to management, from engineering to ecology.

Positive and Negative Feedback

In a system where a transformation occurs, there are *inputs* and *outputs.* The inputs are the result of the environment's influence on the system, and the outputs are the influence of the system on the environment. Input and output are separated by a duration of time, as in before and after, or past and present (Fig. 52).

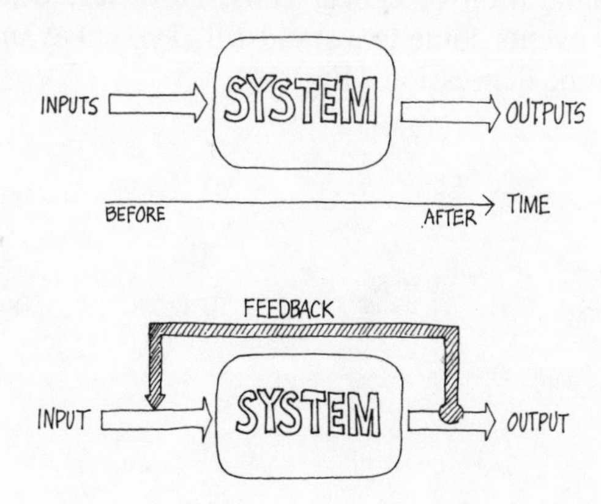

Fig. 52

In every feedback loop, as the name suggests, information about the result of a transformation or an action is *sent back to the input* of the system in the form of input data. If these new data facilitate and accelerate the transformation in the same direction as the preceding results, they are positive feedback—their effects are cumulative. If the new data produce a result in the opposite direction to previous results, they are negative feedback—their effects stabilize the system. In the first case there is exponential growth or decline; in the second there is maintenance of the equilibrium (Fig. 53).

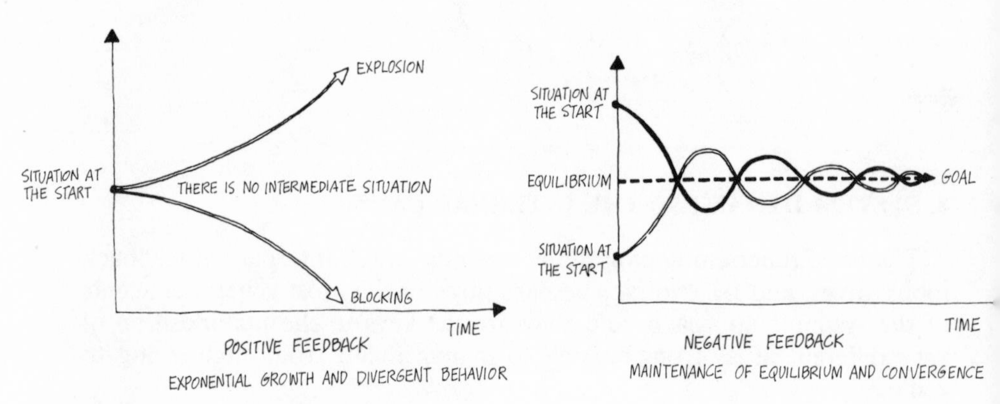

Fig. 53

Positive feedback leads to divergent behavior: indefinite expansion or explosion (a running away toward infinity) or total blocking of activities (a running away toward zero). Each plus involves another plus; there is a snowball effect. The examples are numerous: chain reaction, population explosion, industrial expansion, capital invested at compound interest, inflation, proliferation of cancer cells. However, when minus leads to another minus, events come to a standstill. Typical examples are bankruptcy and economic depression (Fig. 54).

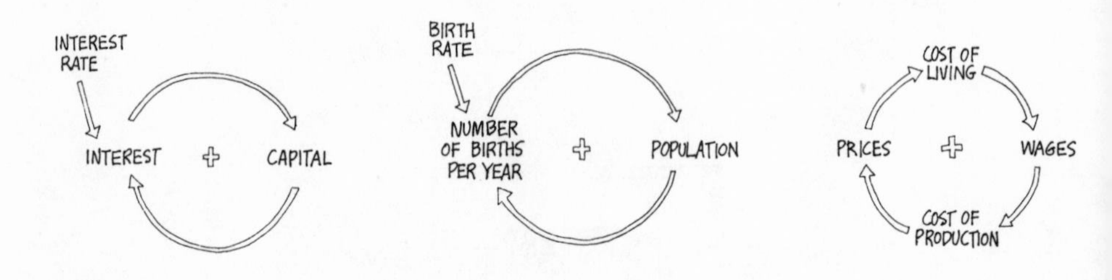

Fig. 54

In either case a positive feedback loop left to itself can lead only to the destruction of the system, through explosion or through the blocking of all its functions. The wild behavior of positive loops—a veritable death wish—must be controlled by negative loops. This control is essential for a system to maintain itself in the course of time.

Negative feedback leads to adaptive, or goal-seeking behavior: sustaining the same level, temperature, concentration, speed, direction. In some cases the goal is self-determined and is preserved in the face of evolution: the system has produced its own purpose (to maintain, for example, the composition of the air or the oceans in the ecosystem or the concentration of glucose in the blood). In other cases man has determined the goals of the machines (automats and servomechanisms).

In a negative loop every variation toward a plus triggers a correction toward the minus, and vice versa. There is tight control; the system oscillates around an ideal equilibrium that it never attains. A thermostat or a water tank equipped with a float are simple examples of regulation by negative feedback (Fig. 55).*

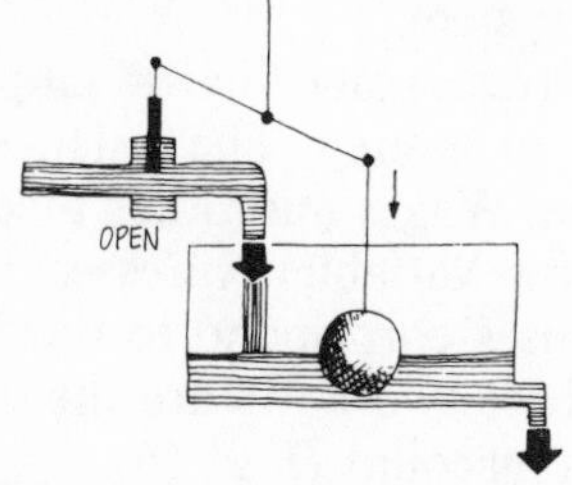

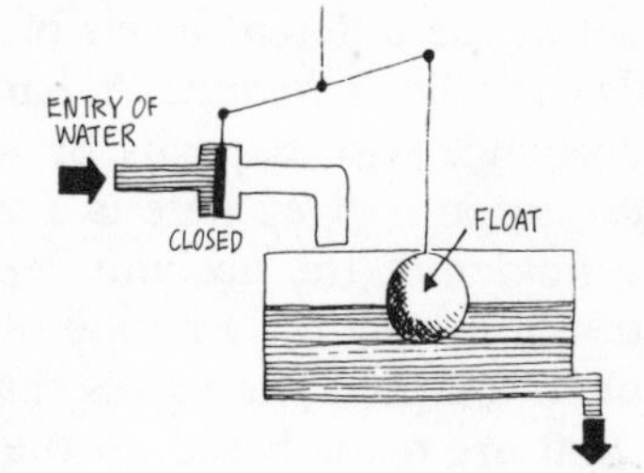

Fig. 55

Flows and Reservoirs

The dynamic behavior of every system, regardless of its complexity, depends in the last analysis on two kinds of variables: flow variables and state or level variables. The first are symbolized by the valves that control the flows, the second (showing what is contained in the reservoirs) by rectangles. The flow variables are expressed only in terms of two instants, or in relation to a given period, and thus are basically functions of time. The state (level) variables indicate the accumulation of a given quantity in the course of time; they express the result of an integration. If time stops, the level remains constant (static level) while the flows disappear—for they are the result of actions, the activities of the system.

Hydraulic examples are the easiest to understand. The flow variable

* Other examples include the controls of population, prices, and the balance of payments (see pp. 10, 25, 26).

is represented by the flow rate, that is, the average quantity running off between two instants. The state variable is the quantity of water accumulated in the reservoir at a given time. If you replace the flow of water by a flow of people (number of births per year), the state variable becomes the population at a given moment.

The difference between flow variables and state variables is illustrated perfectly by the difference between the profit and loss statement and the balance sheet of a firm. The profit and loss statement is concerned with the period between two instants, say January 1 and December 31. It consists of an aggregation of flow variables: salaries paid, total purchases, transportation, interest costs, total sales. The balance sheet applies to one date only, say December 31. It is an instant picture of the situation of the company at that single moment in time. The balance sheet contains a variety of state variables: on the assets side, real estate and property, inventory, accounts receivable; on the liabilities side, capital, long-term debt, accounts payable.

Three examples will serve to explain the relationships between flow variables and state variables and will clarify several of the ways in which they act at the different levels of a complex system.

Balancing one's budget. A bank account (reservoir) fills or empties in accordance with deposits or withdrawals of money. The balance in the account at a given date is a state variable. Wages and other income of the holder of the account represent a flow variable expressed in a quantity of money for a period of time; expenses correspond to the flow variable of output. The valves that control these two flows are the decisions that are made based on the state of the account (Fig. 56).

"To make ends meet" is to establish an equilibrium of the flows: income (input) equal to expenses (output). The bank account is kept at a stationary level. This is a case of dynamic equilibrium.*

When the input flow is greater than the output flow, money accumulates in the account. The owner of the account is "saving." In saving he increases his total income by the amount of interest his savings earn (an example of a positive feedback loop).

When the output flow is greater than the input flow, debts are accumulated. This situation can deteriorate further, for interest on debts increases output (a positive feedback loop toward zero). If the situation is not remedied, it can lead to the exhaustion of funds in a short time.

The maintenance of equilibrium requires tight control. Control can be exercised more easily on the output flow valve (expenses) than on the input flow valve (income). This control imposes a choice of new constraints: the reduction or the better distribution of expenditures. In

---

* This state of equilibrium is accomplished even though the account is emptied and refilled every month. (One could assume that in effect wages were being paid and deposited daily.)

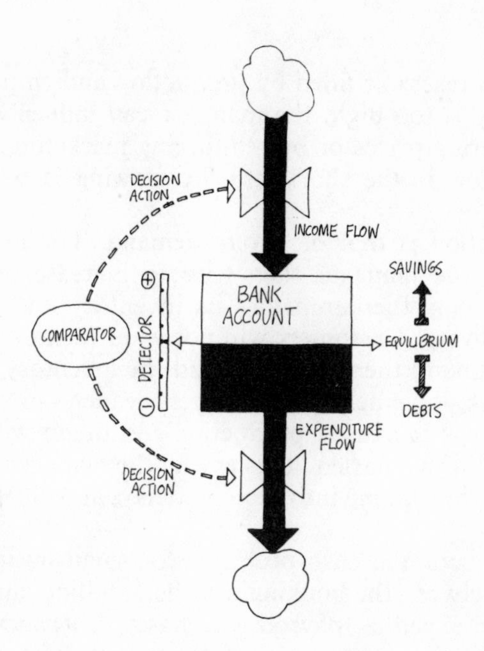

Fig. 56

contrast, to make one's income increase rapidly one has to have reserves (savings) at his disposal—or benefit by a raise in salary.

Managing a company. In the short term the manager uses internal indicators such as sales, inventory, orders placed, changes in production margins, productivity, delivery delays, money in reserve. For longer periods he consults his balance sheet, profit and loss statement, and such outside indicators as the prime rate of interest, manpower, growth of the economy. Using the differences between these indicators and the business forecasts, the manager takes what corrective measures are necessary. Consider two examples related to inventory and cash management.

Fig. 57

An inventory is a reservoir filled by production and emptied by sales. When the inventory is too high, the manager can influence the flow of sales either by lowering prices or by reinforcing marketing. He can also control the input flow in the short term by slowing down production (Fig. 57).

The reverse situation is that of strong demand. The inventory level drops rapidly, and the manager then tries to increase production. If demand remains strong, the company—its inventory low—will require longer delays in delivery. Customers will not want to wait and will turn to a competitor. Demand then decreases and the inventory level climbs. A negative feedback loop helps the business leader—or works to his disadvantage if he has increased production too much without having foreseen the change in the market. This is why the manager must control flow and inventory while taking into account delays and different response times.

One of the most common cash problems for small businesses results from the time lag between the booking of orders, billing, and the receipt of payment. Regular expenses (payroll, purchases, rent) and the irregular receipt of customers' payments together create cash fluctuations. These are eased somewhat by the overdraft privilege that banks grant to some companies. The overdraft exercises a regulatory role like that of inventories, a full backlog of orders, or other reserves: it is the buffering effect we have already encountered, notably in the case of the great reservoirs of the ecological cycle.

Food and world population. Two major variables measure world growth: *industrial capital and population.* The reservoir of industrial capital (factories, machines, vehicles, equipment) is filled through investment and emptied through depreciation, obsolescence, and wear and tear on machines and equipment. The population reservoir is filled by births and emptied by deaths (Fig. 58).

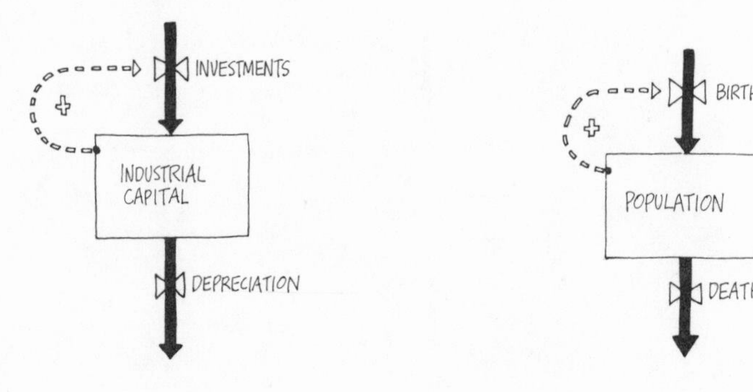

Fig. 58

If the flow of investment is equal to the flow of depreciation, or if births equal deaths, a state of dynamic equilibrium is achieved—a stationary (not static) state called *zero growth.* What will happen then when several flow and state variables interact?

Consider a simple model, the well-known Malthusian model described in classic form. World resources of food grow at a constant rate (a linear, arithmetic progression), while world population grows at a rate that is itself a function of population (a nonlinear, geometric progression) (Fig. 59).

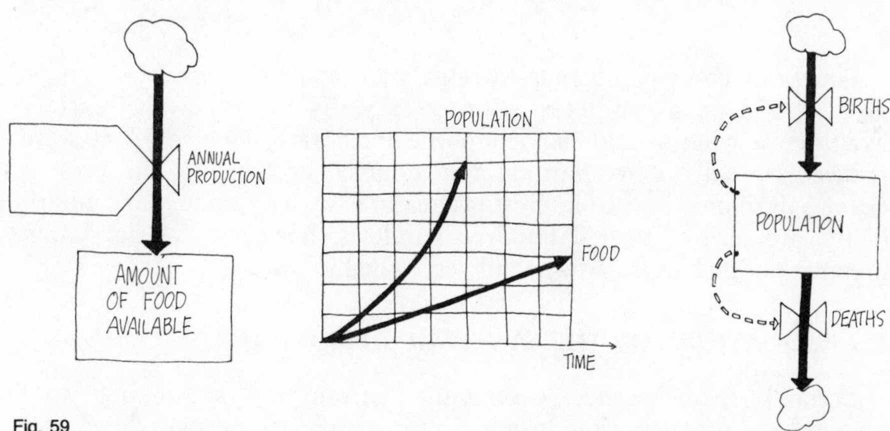

Fig. 59

The food reservoir fills at a constant rate, the population reservoir at an accelerated rate. The control element is represented by the quantity of food available to each individual (Fig. 60).

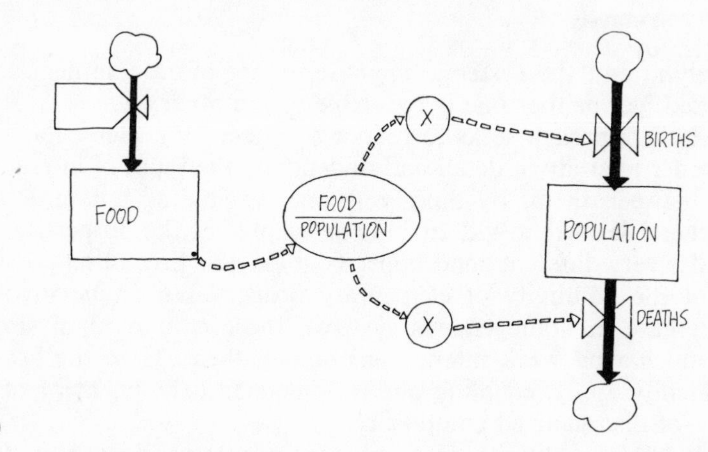

Fig. 60

A decrease in the food quota per person leads to famine and eventually an increase in mortality. The demographic curve stabilizes in an S curve, typical of growth limited by an outside factor (Fig. 61).

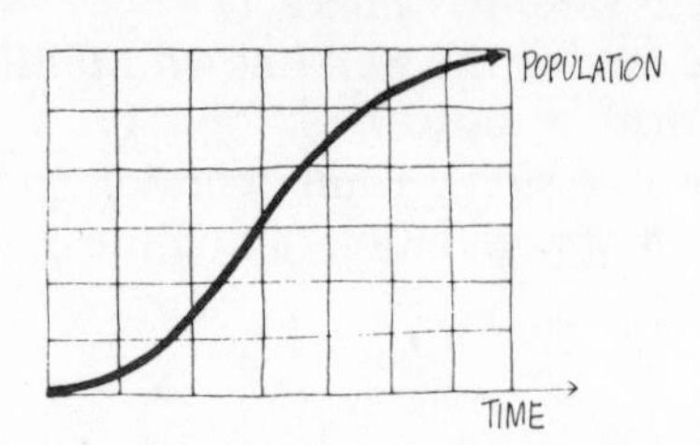

Fig. 61

Equations corresponding to various state and flow variables can be programmed on a computer in order to verify the validity of certain hypotheses: what would happen if the birth rate doubled? if it were reduced by half? if food production doubled or tripled? The present example is of only limited interest because it is such a rudimentary model; in the presence of several hundred variables, however, the simulation presents and achieves, as we shall see, valuable results.

4. APPLICATIONS OF THE SYSTEMIC APPROACH

Certainly there has been a revolution in our way of thinking; what now are the practical uses to which we can put it? Beyond the simple description of the systems of nature it leads to new methods and rules of action—nothing less, as you will see, than the instruction manual for the macroscope.

Analysis and Synthesis

The analytic and the systemic approaches are more complementary than opposed, yet neither one is reducible to the other.

The analytic approach seeks to reduce a system to its elementary elements in order to study in detail and understand the types of interaction that exist between them. By modifying one variable at a time, it tries to infer general laws that will enable one to predict the properties of a system under very different conditions. To make this prediction possible, the laws of the additivity of elementary properties must be invoked. This is the case in homogeneous systems, those composed of similar elements and having weak interactions among them. Here the laws of statistics readily apply, enabling one to understand the behavior of the multitude—of disorganized complexity.

The laws of the additivity of elementary properties do not apply in highly complex systems composed of a large diversity of elements linked together by strong interactions. These systems must be approached by new methods such as those which the systemic approach groups together. The purpose of the new methods is to consider a system in its *totality,* its *complexity,* and its own *dynamics.* Through simulation one can "animate" a system and observe in real time the effects of the different kinds of interactions among its elements. The study of this behavior leads in time to the determination of rules that can modify the system or design other systems.

The following table compares, one by one, the traits of the two approaches.

Analytic Approach	*Systemic Approach*
• isolates, then concentrates on the elements	• unifies and concentrates on the interaction between elements
• studies the nature of interaction	• studies the effects of interactions
• emphasizes the precision of details	• emphasizes global perception
• modifies one variable at a time	• modifies groups of variables simultaneously
• remains independent of duration of time; the phenomena considered are reversible.	• integrates duration of time and irreversibility
• validates facts by means of experimental proof within the body of a theory	• validates facts through comparison of the behavior of the model with reality
• uses precise and detailed models that are less useful in actual operation (example: econometric models)	• uses models that are insufficiently rigorous to be used as bases of knowledge but are useful in decision and action (example: models of the Club of Rome)
• has an efficient approach when interactions are linear and weak	• has an efficient approach when interactions are nonlinear and strong
• leads to discipline-oriented (juxtadisciplinary) education	• leads to multidisciplinary education
• leads to action programmed in detail	• leads to action through objectives
• possesses knowledge of details, poorly defined goals	• possesses knowledge of goals, fuzzy details

This table, while useful in its simplicity, is nevertheless a caricature of reality. The presentation is excessively dualist; it confines thought to an alternative from which it seems difficult to escape. Numerous other points of comparison deserve to be mentioned. Yet without being exhaustive the table has the advantage of effectively opposing the two complementary approaches, one of which—the analytic approach—has been favored disproportionately in our educational system.

To the opposition of analytic and systemic we must add the opposition of static vision and dynamic vision.

Our knowledge of nature and the major scientific laws rests on what I shall call "classic thought," which has three main characteristics.

Its concepts have been shaped in the image of a "solid" (conservation of form, preservation of volume, effects of force, spatial relations, hardness, solidity).
Irreversible time, that of life's duration, of the nondetermined, of chance events, is never taken into account. All that counts is physical time and reversible phenomena. T can be changed to −T without modifying the phenomena under study.
The only form of explanation of phenomena is linear causality; that is, the method of explanation relies on a logical sequence of cause and effect that extends for its full dimension along the arrow of time.

In present modes of thought influenced by the systemic approach, the concept of the fluid replaces that of the solid. Movement replaces permanence. Flexibility and adaptability replace rigidity and stability. The concepts of flow and flow equilibrium are added to those of force and force equilibrium. Duration and irreversibility enter as basic dimensions in the nature of phenomena. Causality becomes circular and opens up to finality.*

The dynamics of systems shatters the static vision of organizations and structures; by integrating time it makes manifest *relatedness* and *development.*

Another table may help to enlighten and enrich the most important concepts related to classic thought and systemic thought (Fig. 62).

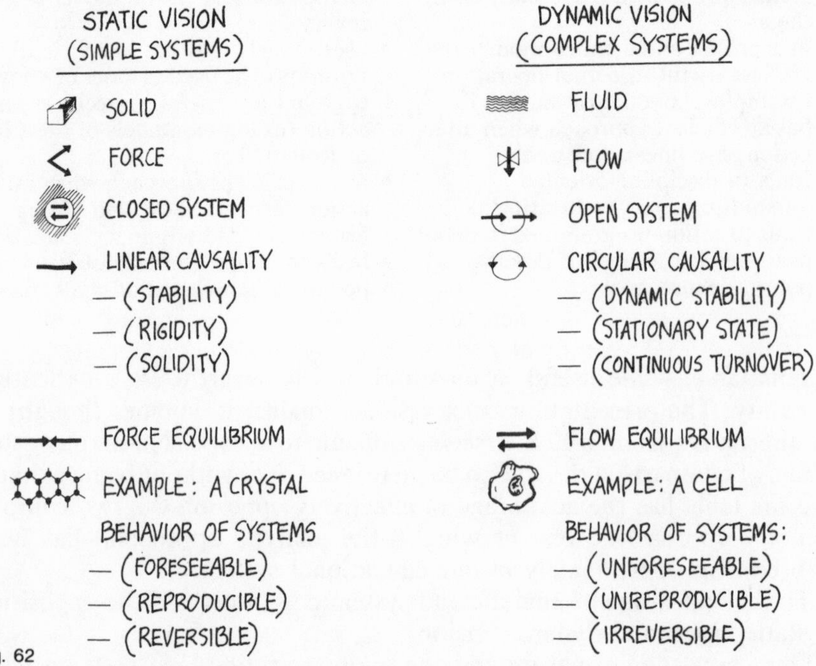

STATIC VISION (SIMPLE SYSTEMS)	DYNAMIC VISION (COMPLEX SYSTEMS)
SOLID	FLUID
FORCE	FLOW
CLOSED SYSTEM	OPEN SYSTEM
LINEAR CAUSALITY — (STABILITY) — (RIGIDITY) — (SOLIDITY)	CIRCULAR CAUSALITY — (DYNAMIC STABILITY) — (STATIONARY STATE) — (CONTINUOUS TURNOVER)
FORCE EQUILIBRIUM	FLOW EQUILIBRIUM
EXAMPLE: A CRYSTAL	EXAMPLE: A CELL
BEHAVIOR OF SYSTEMS — (FORESEEABLE) — (REPRODUCIBLE) — (REVERSIBLE)	BEHAVIOR OF SYSTEMS: — (UNFORESEEABLE) — (UNREPRODUCIBLE) — (IRREVERSIBLE)

Fig. 62

* Numerous points only mentioned here will be taken up again in the following chapters.

Models and Simulation

The construction of models and simulations are among the most widely used methods of the systemic approach, to the extent that they are often confused with the systemic approach itself.

Confronted with complexity and interdependence, we all use simple analogical models. These models, established as part of an earlier analytical approach, seek to unite the main elements of a system in order to permit hypotheses concerning the behavior of the system as a whole—by taking into account as much as possible the interdependence of the factors.

When the number of variables is small, we constantly use such analogical models to understand a system of which we have little information or to try to anticipate the responses or reactions of someone with a different model of the situation. Our vision of the world is a model. Every mental image is a fuzzy, incomplete model that serves as a basis for decision.

The construction of simple analogical models rapidly becomes impracticable when large numbers of variables are involved. This is the case with highly complex systems. The limitations of our brain make it impossible for us to make a system "live" without the help of computers and simulation systems, so we turn to these mechanical and electronic means.

Simulation tries to make a system live by simultaneously involving all its variables. It relies on a model established on the basis of previous analysis. Systems analysis, model building, and simulation are the three fundamental stages in the study of the dynamic behavior of complex systems.

Systems analysis defines the limits of the system to be modeled, identifies the important elements and the types of interactions between these elements, and determines the connections that integrate the elements into an organized whole. Elements and types of connections are classified and placed in hierarchical order. One may then extract and identify the flow variables, the state variables, positive and negative feedback loops, delays, sources, and sinks. Each loop is studied separately, and its influence on the behavior of the different component units of the system is evaluated.

Model building involves the construction of a model from data provided by systems analysis. One establishes first a complete diagram of the causal relations between the elements of the subsystem. (In the Malthusian model on p. 77 these include the influences of birth rate on population and food rationing on mortality.) Then, in the appropriate computer language, one prepares the equations describing the interactions and connections between the different elements of the system.

Simulation considers the dynamic behavior of a complex system. Instead of modifying one variable at a time it uses a computer to set in motion simultaneously groups of variables in order to produce a real-life situation. A simulator, which is an interactive physical model, can also be used to give in "real time" the answers to different decisions and reactions of its user. One such simulator is the flight simulator used by student pilots. Simulation is used today in many areas, thanks to the development of more powerful yet simpler simulation language and new interactive means of communication with the computer (graphic output on cathode ray tubes, high-speed plotters, input light pens, computer-generated animated films).

Examples of the applications of simulation are to be found in many fields: *economy and politics*—technological forecasting, simulation of conflicts, "world models"; *industrial management*—marketing policy, market penetration, launching a new product: *ecology*—effects of atmospheric pollutants, concentration of pollutants in the food chain; *city planning*—growth of cities, appearance of slums, automobile traffic; *astrophysics*—birth and evolution of the galaxies, "experiments" produced in the atmosphere of a distant planet; *physics*—the flow of electrons in a semiconductor, resistance of materials, shock waves, flow of liquids, formation of waves; *public works*—silting-in of ports, effects of wind on high-rise buildings;*chemistry*—simulation of chemical reactions, studies of molecular structure; *biology*—circulation in the capillaries, competitive growth between bacterial populations, effects of drugs, population genetics; *data processing*—simulation of the function of a computer before its construction; *operational research*—problems of waiting lines, optimization, resource allocation, manufacturing control; *engineering*—process control, calculations of energy costs, calculations of construction costs; *education*—simulated pedagogical practices, business games.

Despite the number and diversity of these applications, too much cannot be expected of simulation. It is only one approach among many, a complementary method of studying a complex system. Simulation never gives the optimum or the exact solution to a given problem. It only sets forth the general tendencies of the behavior of a system—its probable directions of evolution—while suggesting new hypotheses.

One of the serious dangers of simulation results from too much freedom in the choice of variables. The user can change the initial conditions "just to see what will happen." There is the risk of becoming lost in the infinity of variables and the incoherent performances associated with chance modifications. The results of simulation must not be confused with reality (as is often the case) but, compared with what one knows of reality, should be used as the basis for the possible modification of the initial model. When one continues to use such a process in successive approximations, the usefulness of simulation will become apparent.

Simulation appears to be one of the most resourceful tools of the systemic approach. It enables us to verify the effects of a large number of variables on the overall functioning of a system; it ranks the role of each variable in order of importance; it detects the points of amplification or inhibition through which we can influence the behavior of the system. The user can test different hypotheses without running the risk of destroying the system under study—a particularly important advantage in the case of living systems or those that are fragile or very costly.

Knowing that one can experiment on a model of reality rather than on reality itself, one can influence the time variable by accelerating very slow phenomena (social phenomena, for example) or slowing down ultra-fast phenomena (the impact of a projectile on a surface). One can influence equally well the space variable by simulating the interactions that occur in very confined volumes or over great distances.

Simulation does not bring out of the computer, as if by magic, more than what was put into the program. The contribution of the computer rests at a qualitative level. Handling millions of bits of information in a tiny fraction of time, it reveals structures, modes, and tendencies heretofore unobservable and which result from the dynamic behavior of the system.

Interaction between user and model develops a feeling of the effect of interdependencies and makes it possible to anticipate better the reactions of the models. Evidently this feeling exists for all those who have had long experience in the management of complex organizations. One of the advantages of simulation is that it allows the more rapid acquisition of these fundamental mechanisms.

Finally, simulation is a new aid to decision making. It enables one to make choices among "possible futures." Applied to social systems, it is not directly predictive. (How does one take into account such impossible-to-quantify data as well-being, fear, desire, or affective reactions?) Yet simulation does constitute a sort of portable sociological laboratory with which experiments can be made without involving the future of millions of men and without using up important resources in programs that often lead to failure.

Certainly models are still imperfect. As Dennis Meadows observed, however, the only alternatives are "mental models" made from fragments of elements and intuitive thinking. Major political decisions usually rest on such mental models.

The Dynamics of Maintenance and Change

The properties and the behavior of a complex system are determined by its internal organization and its relations with its environment. To understand better these properties and to anticipate better its behavior, it is necessary to act on the system by transforming it or by orienting its evolution.

Every system has two fundamental modes of existence and behavior: *maintenance* and *change.* The first, based on negative feedback loops, is characterized by *stability.* The second, based on positive feedback loops, is characterized by *growth* (or *decline.*) The coexistence of the two modes at the heart of an open system, constantly subject to random disturbances from its environment, creates a series of common behavior patterns. The principal patterns can be summarized in a series of simple graphs by taking as a variable any typical parameter of the system (size, output, total sales, number of elements) as a function of time (Fig. 63).*

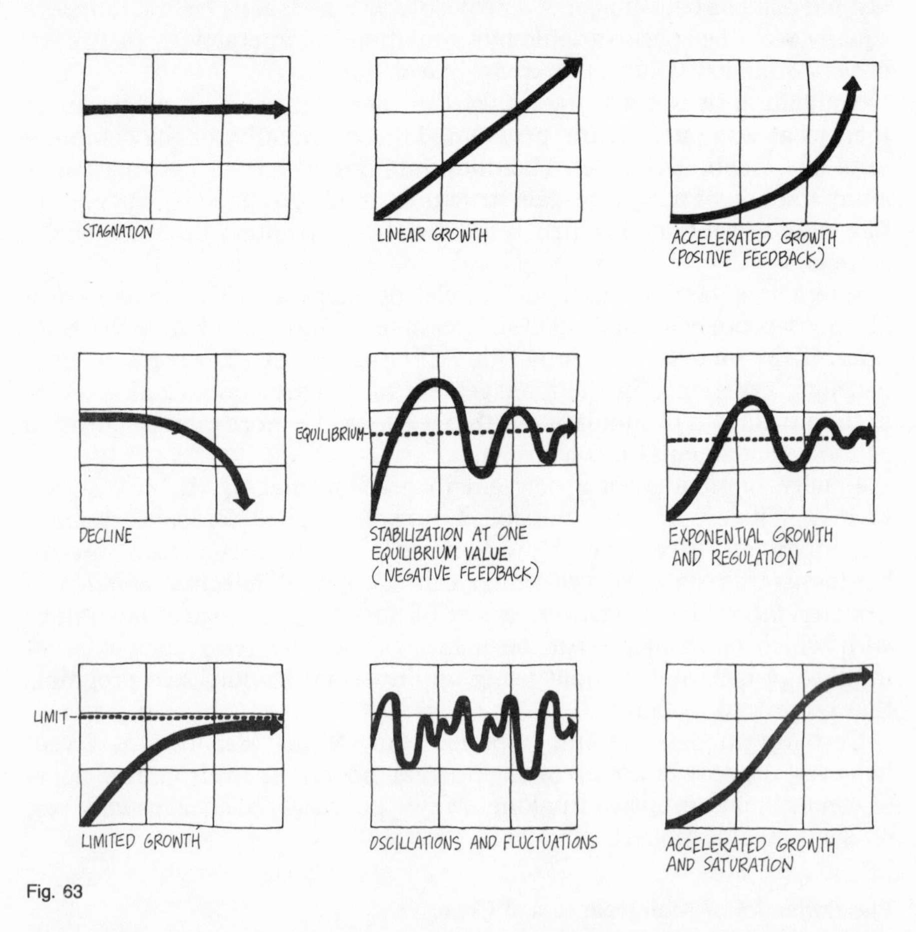

Fig. 63

* It must be remembered that the overall behavior of a system is the result of the individual behavior patterns of subsystems, patterns themselves determined by the interconnection of a large number of variables.

Dynamic stability: equilibrium in motion. Maintenance is duration. Negative controls, by regulating the divergences of positive loops, help to stabilize a system and enable it to last. Thus the system is capable of self-regulation.

To bring together stability and dynamics might seem to be paradoxical. In fact the juxtaposition demonstrates that the structures or the functions of an open system remain identical to themselves in spite of the continuous turnover of the components of the system. This persistence of form is dynamic stability. It is found in the cell, in the living organism, in the flame of a candle.

Dynamic stability results from the combination and readjustment of numerous equilibriums attained and maintained by the system—that of the "internal milieu" of the living organism, for example (see p. 42). We deal with dynamic equilibriums; this imposes a preliminary distinction between balance of force and balance of flow.

A *balance of force* results from the neutralization at the same point of two or more equal and opposed forces. This might be illustrated by two hands immobilized in handcuffs or by a ball lying in the bottom of a basin (Fig. 64).

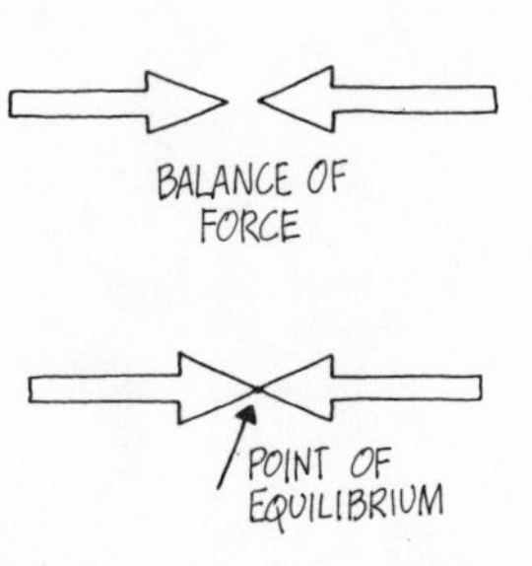

Fig. 64

When there are two forces present—two armies or two governments—we speak of the "balance of power." But a balance of force is a static equilibrium; it can be modified only as the result of a discontinuous change in the relationship of the forces. This discontinuity could lead to an escalation when one force overpowers the other.

On the other hand, a *balance of flow* results from the adjustment of the speeds of two or more flows crossing a measuring device. Equilibrium exists when the speeds of the flows are equal and moving in opposite directions.* This is the case of a transaction at a sales counter, where merchandise is exchanged for money (Fig. 65).

---

* Or when flows have opposite effects even though they move in the same direction (as a reservoir filling and emptying at the same time).

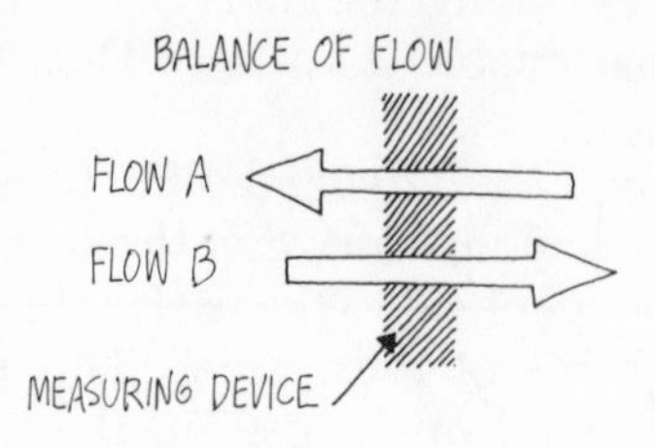

Fig. 65

A balance of flow is a dynamic equilibrium. It can be adapted, modified, and modeled permanently by often imperceptible readjustments, depending on disturbances or circumstances. The balance of flow is the foundation of dynamic stability.

When equilibrium is achieved, a given "level" is maintained over time (like the concentration of certain molecules in the plasma, or the state of a bank account—see p. 74). This particular state is called a *steady state;* it is very different from the *static state* represented by the level of water in a reservoir having no communication with the environment (Fig. 66).

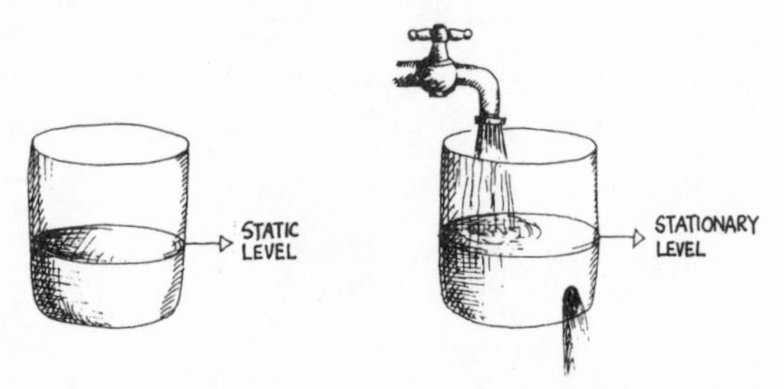

Fig. 66

There are as many steady states as there are levels of equilibrium at different depths of a reservoir. This makes it possible for an open system to adapt and respond to the great variety of modifications in the environment.

Homeostasis: resistance to change. Homeostasis is one of the most remarkable and most typical properties of highly complex open systems.

The term was created by the American physiologist Walter Cannon in 1932 (see p. 43). A homeostatic system (an industrial firm, a large organization, a cell) is an open system that maintains its structure and functions by means of a multiplicity of dynamic equilibriums rigorously controlled by interdependent regulation mechanisms. Such a system reacts to every change in the environment, or to every random disturbance, through a series of modifications of equal size and opposite direction to those that created the disturbance. The goal of these modifications is to maintain the internal balances.

Ecological, biological, and social systems are homeostatic. They oppose change with every means at their disposal. If the system does not succeed in reestablishing its equilibriums, it enters into another mode of behavior, one with constraints often more severe than the previous ones. This mode can lead to the destruction of the system if the disturbances persist.

Complex systems must have homeostasis to maintain stability and to survive. At the same time it bestows on the systems very special properties. Homeostatic systems are ultrastable; everything in their internal, structural, and functional organization contributes to the maintenance of the same organization. Their behavior is unpredictable; "counterintuitive" according to Jay Forrester, or contravariant: when one expects a determined reaction as the result of a precise action, a completely unexpected and often contrary action occurs instead. These are the gambles of interdependence and homeostasis; statesmen, business leaders, and sociologists know the effects only too well.

For a complex system, to endure is not enough; it must adapt itself to modifications of the environment and it must evolve. Otherwise outside forces will soon disorganize and destroy it. The paradoxical situation that confronts all those responsible for the maintenance and evolution of a complex system, whether the system be a state, a large organization, or an industry, can be expressed in the simple question, How can a stable organization whose goal is to maintain itself and endure be able to change and evolve?

Growth and variety. The growth of a complex system—growth in volume, size, number of elements—depends on positive feedback loops and the storage of energy. In effect a positive feedback loop always acting in the same direction leads to the accelerated growth of a given value (see p. 73). This value can be *number* (population growth), *diversity* (variety of elements and interactions between elements), or *energy* (energy surplus, accumulation of profits, growth of capital).

The positive feedback loop is equivalent to a *random variety generator.* It amplifies the slightest variation; it increases the possibilities of choice, accentuates differentiation, and generates complexity by increasing the possibilities for interaction.

Variety and complexity are closely allied. Variety, however, is one of the conditions for the stability of a system. In fact homeostasis can be established and maintained only when there is a large variety of controls. The more complex a system, the more complex its control system must be in order to provide a "response" to the multiple disturbances produced by the environment. This is the *law of requisite variety* proposed by Ross Ashby in 1956. This very general law asserts in mathematical form that the regulation of a system is efficient only when it depends on a system of controls *as complex as the system itself.* In other words, control actions must have a variety equal to the variety of the system. In ecology, for example, it is the variety of species, the number of ecological niches, the abundance of interactions among species and between community and environment that guarantee the stability and continuance of the community. Variety permits a wider range of response to potential forms of aggression from the environment.

The generation of variety can lead to adaptations through increase in complexity. But in its confrontation with the random disturbances of the environment, variety also produces the *unexpected,* which is the seed of change. Growth is then both a force for change and a means for adapting to the modifications of the environment. Here one begins to see the way in which a homeostatic system can evolve as a system constructed to resist change. It evolves through a complementary process of total or partial disorganization and reorganization. This process is produced either by the confrontation of the system with random disturbances from the environment (mutations, events, "noise") or in the course of readjustment of an imbalance (resulting, for example, from too rapid growth).

Evolution and emergence. Living systems can adapt, within certain limits, to sudden modifications coming from the outside world. A system actually has detectors and comparators that enable it to detect signals from within or without and to compare the signals to equilibrium values. When there are discrepancies, the emission of error signals can help to correct them. If it cannot return to its former state of homeostatic equilibrium, the system, through the complementary play of positive and negative feedback loops, searches for new points of equilibrium and new stationary states.

The evolution of an open system is the integration of these changes and adaptations, the accumulation in time of successive plans or "layers" of its history.* This evolution materializes through hierarchical levels of organization and the emergence of new properties. The prebiological evolution (the genesis of living systems) and the biological and social evolutions are examples of evolution toward levels of increasing complexity. At each level new properties "emerge" that cannot be explained by the sum of the properties of each of the parts which constitute the whole. There is a qualitative leap; the crossing of a threshold; life, reflective thought, and collective consciousness.

Emergence is linked to complexity. The increase in the diversity of elements, in the number of connections between these elements, and in the play of nonlinear interactions leads to patterns of behavior that are

* Mechanisms of evolution will be introduced in the fifth chapter.

difficult to predict—especially if they are founded solely on the properties of the elements. We know, for example, the properties of each of the amino acids that make up the protein chain. But because of the convolutions of this chain, certain amino acids that are far apart in the sequence find themselves together in space. This situation gives the protein emergent properties that enable it to recognize certain molecules and to catalyze their transformation. This would be impossible if the amino acids were present in the milieu but not arranged in the proper order—or if the chain were straightened out.

The "Ten Commandments" of the Systemic Approach

The systemic approach has little value if it does not lead to practical applications such as facilitating the acquisition of knowledge and improving the effectiveness of our actions. It should enable us to extract from the properties and the behavior of complex systems some general rules for understanding systems better and acting on them.

Unlike the juridical, moral, or even physiological laws which one might still cheat, a misappreciation of some of the basic systemic laws could result in serious error and perhaps lead to the destruction of the system within which one is trying to act. Of course many people will have an intuitive knowledge of these laws, which are very much the result of experience or simple common sense. The following are the "ten commandments" of the systemic approach.

1. Preserve variety. To preserve stability one must preserve variety. Any simplification is dangerous because it introduces imbalance. Examples abound in ecology. The disappearance of some species as a consequence of the encroaching progress of "civilization" brings the degradation of the entire ecosystem. In some areas intensive agriculture destroys the equilibrium of the ecological pyramid and replaces it with an unstable equilibrium of only three stages (grain, cattle, and man) controlled by a single dominant species. This unbalanced ecosystem tries spontaneously to return to a state of higher complexity through the proliferation of insects and weeds—which farmers prevent by the widespread use of pesticides and herbicides.

In economy and in management, excessive centralization produces a simplification of communication networks and the impoverishment of the interactions between individuals. There follow disorder, imbalance, and a failure to adapt to rapidly changing situations.

2. Do not "open" regulatory loops. The isolation of one factor leads to prompt actions, the effects of which often disrupt the entire system. To obtain a short-term action, a stabilizing loop or an overlapping series of feedback loops is often "cut open"—in the belief that one is acting directly on the causes in order to control the effects. This is the cause of sometimes dramatic errors in medicine, economy, and ecology.

Consider some examples of what happens in the rupture of natural cycles. The massive use of fossil fuels, chemical fertilizers, or nonrecyclable pesticides allows the agricultural yield to grow in the short term; in the long term this action may bring on irreversible disturbances. The fight against insects leads as well to the disappearance of the birds that feed on the insects; the result in the long term is that the insects return in full force—but there are no birds. The states of waking, sleeping, and dreaming are probably regulated by the delicate balance between chemical substances that exist in the brain; by regularly introducing, for short-term effect, an outside foreign molecule such as a sleeping pill, the natural long-term mechanisms are inhibited—worse, there is the danger of upsetting them almost irrevocably: persons accustomed to using barbiturates must undergo a veritable detoxification in order to return to a normal sleep pattern.

3. Look for the points of amplification. Systems analysis and simulation bring out the *sensitive points* of a complex system. By acting at this level, one releases either amplifications or controlled inhibitions.

A homeostatic system resists every measure, immediate or sequential (that is, waiting for the results of preceding measures in order to take on new ones). One of the methods that influence the system and cause it to evolve in a chosen direction is the use of a *policy mix.* These measures must be carefully proportioned in their relationships and applied simultaneously at different points of influence.

One example is the problem of solid wastes. There are only three ways to reduce the flow of the generation of solid wastes by acting on the valve (the flow variable): reducing the number of products used (which would mean a drop in the standard of living), reducing the quantity of solid wastes in each product, or increasing the life expectancy of the products by making them more durable and easier to repair. The simulations performed by Jorgan Randers of MIT show that no one measure alone is enough. The best results came from a policy mix, a combination of measures used at the same time: a tax of 25 percent on the extraction of nonrenewable resources, a subsidy of 25 percent for recycling, a 50 percent increase in the life of the products, a doubling of the recyclable portion per product, and a reduction in primary raw material per product (Fig. 67).

4. Reestablish equilibriums through decentralization. The rapid reestablishment of equilibriums requires the detection of variances where they occur and corrective action that is carried out in a decentralized manner.

The correction of the body's equilibrium when we stand is accomplished by the contraction of certain muscles without our having to

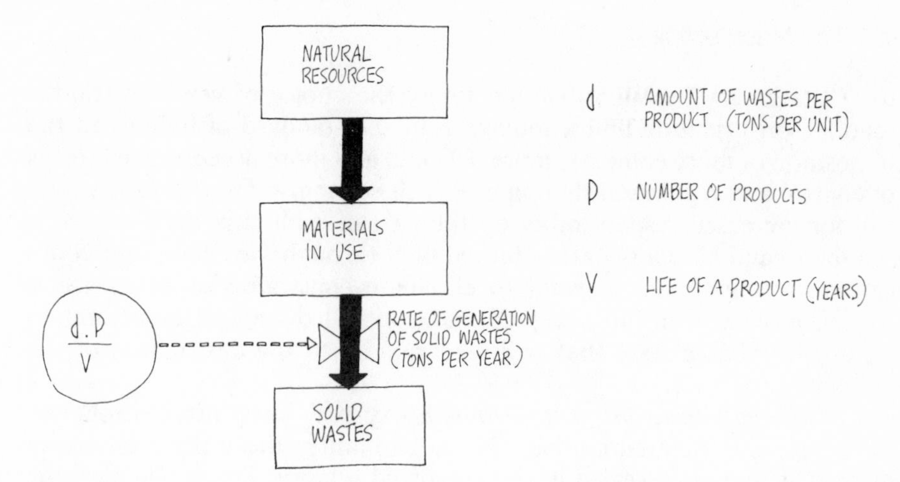

Fig. 67

think about it even when the brain intervenes. Enzymatic regulation networks show that the entire hierarchy of levels of complexity intervene in the reestablishment of balance (recall the example of the service station, p. 51). Often corrective action has been taken even before one has been made conscious of taking it. The decentralization of the reestablishment of equilibriums is one application of the law of requisite variety. It is customary in the body, the cell, the ecosystem. But so far it appears that we have not succeeded in applying this law to the organizations that we have been assigned to manage.

5. Know how to maintain constraints. A complex open system can function according to different modes of behavior. Some of them are desirable; others lead to the disorganization of the system. If we want to maintain a given behavior that we consider preferable to another, we must accept and maintain certain kinds of constraints in order to keep the system from turning toward a less desirable or a dangerous mode of behavior.

In the management of the family budget one can choose a high style of living (living beyond one's means), with the constraints that it implies with respect to banks and creditors. Or one can choose to limit expenditures and do without goods one would like to possess—a different set of constraints.

In the case of a nation's economy, those responsible for the economic policy choose and maintain the constraints that result from inflation, with all their injustices and social inequalities—for they are judged a lesser evil than those brought about by unemployment.

At the level of the world economy the growth race entails social inequalities, depletion of resources, and pollution. Theoretically, however, it allows a more rapid increase in the standard of living. The transition to a "stationary" economy would imply the choice of new constraints, founded on privation and a reduction in the standard of living and the imposition of more complex, more delicate, and more decentralized forms of control and regulation than in a growth economy. These means would call for increased responsibility on the part of each citizen.

Liberty and autonomy are achieved only through the choice and application of constraints; to want to eliminate constraints at any price is to risk moving from an accepted and controlled state of constraint to an uncontrollable state that will lead rapidly to the destruction of the system.

6. Differentiate to integrate better. Every real integration is founded on a previous differentiation. The individuality, the unique character of each element is revealed in the organized totality. This is the meaning of Teilhard de Chardin's famous phrase, "union differentiates." This law of the "personalizing" union is illustrated by the specialization of cells in the tissues or the organs of the body.

There is no true union without antagonism, balance of power, conflict. Homogeneity, mixture, and syncretism are forms of entropy. Only union through diversity is creative; it increases complexity and leads to higher levels of organization. This systemic law and its allied constraints are well known by those whose purpose is to unite, to assemble, to federate. Antagonism and conflict are always born of the transition to a unified entity. Before regrouping diversities, we must decide to what limits we should push the process of personalization. Pushed too soon, it leads to an homogenizing and paralyzing mixture; pushed too late, it leads to the confrontation of individualism and personality—and perhaps a disassociation still greater than what had formerly existed.

7. To evolve, allow aggression. A homeostatic (ultrastable) system can evolve only if it is assaulted by events from the world outside. An organization must then be in a position to capture the germs of change and use them in its evolution—which obliges it to adopt a mode of functioning characterized by the renewal of structures and the mobility of men and ideas. In effect all rigidity, sclerosis, and perpetuity of structures or hierarchy is clearly opposed to a system that allows evolution.

An organization can maintain itself in the manner of a crystal or that of a living cell. The crystal preserves its structure by means of the balance of forces that cancel out each other in every node of the crystalline network—and by redundancy, or repetition of patterns. This static state, closed to the environment, allows no resistance to change within its milieu: if the temperature rises, the crystal becomes disorganized and melts. The cell, however, is in dynamic equilibrium with its environment. Its organization is founded not on repetition but on the variety of its elements. An open system, it maintains a constant turnover of its elements. Variety and mobility enable it to adapt to change.

The crystal-like organization evolves slowly in the give and take of radical and traumatic reforms. The cell-like organization tries to make the most of events, variety, and the openings into the outside world. It is not afraid of a passing disorganization—the most efficient condition for readaptation. To accept this transitory risk is to accept and to want change. For there is no real change without risk.

8. Prefer objectives to detailed programming. The setting of objectives and rigorous control—as opposed to detailed programming at every step—is what differentiates a servomechanism from a rigidly programmed automatic machine. The programming of the machine must foresee all disturbances likely to occur in the course of operation. The servomechanism, however, adapts to complexity; it needs only to have its goal set without ambiguity and to establish the means of control that will enable it to take corrective measures in the course of action.

These basic principles of cybernetics apply to every human organization. The definition of objectives, the means of attaining them, and the determination of deadlines are more important than the detailed programming of daily activities. Minutely detailed programming runs the risk of being paralyzing; authoritarian programming leaves little room for imagination and involvement. Whatever roads are taken, the important thing is to arrive at the goal—provided that the well-defined limits (necessary resources and total time allotted to operations) are not exceeded.

9. Know how to use operating energy. Data sent out by a command center can be amplified in significant proportions, especially when the data are relayed by the hierarchical structures of organizations or by diffusion networks.

At the energy level the metabolism of the operator of a machine is negligible compared to the power that he can release and control. The same applies to a manager or to anyone in charge of a large organization.

We must distinguish, then, between *power energy* and *operating energy.* Power energy is represented by the electric line or the current that heats a resistance; or it may be the water pipe that carries water pressure to a given point. Operating energy renders itself in the action of the thermostat or the water tap: it represents information.

A servomechanism distributes its own operating energy through the distribution of information that commands its operational parts. In the same way the leader of an organization must help his own system to distribute its operating energy. To accomplish this he establishes feedback loops to the decision centers. In the management of an industry or in the structure of a government, these regulatory loops are called self-management *(autogestion),* participation, or social feedback.*

* Social feedback will be discussed in the fourth chapter.

10. Respect response times. Complex systems integrate time into their organization. Each system has a response time characteristic of that system, by reason of the combined effects of feedback loops, delays at reservoirs, and the sluggishness of flows. In many cases, especially in industry, it is useless to look for speed of execution at any price, to exert pressure in order to obtain responses or results. It is better to try to understand the internal dynamics of the system and to anticipate delays in response. This type of training is often acquired in the actual running of large organizations. It gives rise to *a sense of timing,* the knowing when to begin an action, neither too soon nor too late, but at the precise moment the system is ready to move in one direction or the other. Sense of timing allows the best possible use of the internal energy of a complex system—rather than to have to impose instructions from outside against which the system will react.

Avoiding the Dangers of the Systemic Approach

To be useful, the systemic approach must be demystified; what is useful in daily life must not be reserved for a small elite. The hierarchy of disciplines established in the nineteenth century, from the "most noble" sciences (mathematics and physics) to the "least noble" (the sciences of man and society), continues to weigh heavily on our approach to nature and our vision of the world. Skepticism or distrust of the systemic approach is found among those—mathematicians and physicists—who have received the most advanced theoretical training. At the same time, those who by nature of their research have been accustomed to think in terms of flow, transfer, exchange, and irreversibility—biologists, economists, and ecologists—assimilate more naturally the systemic concepts and communicate more easily among themselves.

To demystify further the systemic approach and to enable it to remain a transdisciplinary *attitude,* a *training* in the mastery of complexity and interdependence, it may be necessary to get rid of the very terms *systemic approach* and *systemic method.* The global vision is not reserved for the few with wide responsibility—the philosophers and the scientists. Each one of us can see things in perspective. We must learn to look through the macroscope to apply systemic rules, to construct more rigorous mental models, and perhaps to master the play of interdependencies.

And we must not hide the dangers of a too systematic use of the systemic approach. A purely descriptive approach—the "what is linked to what?" method—leads rapidly to a collection of useless models of the different systems of nature. The greatest generalization of the concept of system can also turn against itself, destroying its fecundity in sterilizing platitude. In the same way the uncontrolled use of analogies, homologies, and isomorphisms can result in interpretations that complicate rather than enlighten. Such interpretations are founded on superficial resemblances rather than on principles and fundamental laws that are common to all systems. According to Edgar Morin, "too much unification can become abusive simplification, then an *idée fixe* or a turn of phrase."

Once again we are encumbered with the danger of dogmatism. The systemic approach leads to an intransigent systematism or a reductionist biologism. There is danger of our being seduced by models that were conceived as ends of reflective thought, not as points of departure for research. We are tempted by the too simplistic transposition of models or biological laws to society.* The cybernetics of regulation at the molecular level offers general models, some aspects of which are transposable, with certain restrictions, to social systems. The greatest weakness of these models is that they apparently cannot take into account the relationship between force and the conflicts that arise between the elements of every socioeconomic system. The economist J. Attali remarked on this at a meeting of the Group of Ten devoted to the maintenance of biological and social equilibriums: "Unlike the sociologist, the biologist observes systems with well-established laws: they do not change as they are being studied. As for molecules, cells, or microbes, they will never complain of their condition!"

One of the greatest dangers that menace the systemic approach is the temptation of the "unitary theory," the all-inclusive model with all the answers and the ability to predict everything. The use of mathematical language, which by nature and vocation generalizes, can lead to a formalism that isolates the systemic approach instead of opening it up to the practical. The General System Theory does not escape this danger. Sometimes it becomes locked into the language of graph theory, set theory, game theory, or information theory; sometimes it is nothing more than a collection of descriptive approaches that are often illuminating but have no practical application.

The functional systemic approach offers one way of bypassing these alternatives. It avoids the dangerous stumbling blocks of paralyzing reductionism and total systematism; it clears the way for the communication of knowledge, for action, and for creation. For the communication of knowledge because the systemic approach has a conceptual framework

* The danger of too direct transpositions from the biological to the social realms were clearly perceived by Friedrich Engels when he wrote to the Russian sociologist and journalist Piotr Lavrov in 1875: "The essential difference between human society and animal society is that animals, at best, *collect* while men *produce.* This unique but major difference prohibits in its own right the transfer pure and simple of the laws of animal societies to the social systems of men." The work of A. J. Lotka in 1925 on the dynamics of population and the work of V. Volterra in 1931 on the mathematical theory of the life struggle have subsequently shown that we must be less dogmatic than Engels with respect to transfers from the biological to the social realms.

of reference that helps to organize knowledge as it is acquired, reinforces its memorization, and facilitates its transmission. For action because the systemic approach provides rules for confronting complexity and because it assigns to their hierarchical order the elements that are the basis for decisions. And for creation because the systemic approach catalyzes imagination, creativity, and invention. It is the foundation of inventive thought (where the analytical approach is the foundation of knowledgeable thought). Tolerant and pragmatic, systemic thought is open to analogy, metaphor, and model—all formerly excluded from "the scientific method" and now rehabilitated. Everything that unlocks knowledge and frees imagination is welcomed by the systemic approach; it will remain open, like the systems it studies.

The earth shelters the embryo of a body and the beginnings of a spirit. The life of this body is maintained by the great ecological and economic functions brought together in the ecosphere. Collective consciousness emerges from the simultaneous communication of men's brains; it constitutes the noosphere.

Ecosphere and noosphere have energy and information as a base. Action is the synthesis of energy and information. But all action requires time. Thus time is the link between energy, information, and action. The following chapters will be devoted to such a global approach toward energy, information, and time—through trying to envisage old problems from a new perspective.

Three

Energy and Survival

The energy crisis has revealed the physical aspect of human society in sudden and dramatic fashion. Nothing can escape the implacable laws of thermodynamics; human society, like any machine or organism, is no exception. Economists are finding this out now, apparently with some surprise, in the wake of the discoveries of biologists and ecologists.

The necessary tools for considering the overall picture of the flow and degradation of energy in human society—the metabolism of the social organism, its primary function of self-maintenance—have been available for only a short time. Observing this metabolism through the macroscope, we see its dynamic behavior, heretofore impossible to grasp from within.

Out of the relationship between the "anatomy" and the "physiology" of society the long-unsuspected link between energy, economy, ecology, and entropy has been brought to light. This relationship not only reveals the possible causes of the ills of the social organism, it suggests the kinds of remedies that one might apply to a system on which the lives of all of us depend.

It has taken several years for biologists to arrive at an all-encompassing vision of the flow of energy in living systems and to create the new discipline that we now know as *bioenergetics.* Yet most biochemistry textbooks used by medical students have retained the analytical approach, describing in detail the behavior and functions of families of molecules, while the systemic approach considers the universal functioning of the cell. The situation is still worse when one tries to study the ecosphere as a whole. Until now the piecemeal analytical approach alone has prevailed. Going beyond bioenergetics, then, I want to propose the term *ecoenergetics* to show the need for a global approach devoted to the study of the regulation of energy flows in society.

Ecoenergetics must depend on both the systemic and the analytical approaches. With the former we want to proceed to a global study of the transformation and utilization of energy in society. With the latter we want to make a detailed analysis of all transfers of energy that affect the functions of production, consumption, and recovery in the social system. This study is an *energy analysis.*

The purpose of ecoenergetics is to find ways to keep the industrial and economic activities of man from interfering with nature's cycles—above all, to establish bases for real and effective cooperation between man and nature, abandoning forever the old idea of domination. This chapter, then, will address itself to analysis of the crises we face and new proposals that look toward long-term solutions.

1. THE DOMESTICATION OF ENERGY

The story of society is usually told in history books as the political and economic evolution of a country. Yet the laws of energy are also significant. Why not retell the history of social organization from the point of view of energy? The approach is justified because energy laws have priority over political and economic laws. Energy laws are the basis of action; energy is essential to maintenance, to change, and to progress in any organization. Every surplus of energy represents "a leap forward." Prebiological evolution (that which preceded the appearance of the first cells) provides an excellent example.

In the primeval oceans the first organisms found themselves surrounded by energy-rich organic substances that had been accumulating for millions of years. The discovery was like that of man's finding fossil fuels deep in the earth. The energy released through fermentation gave these rudimentary organisms the means to survive, but only in a marginal way and while accumulating toxic wastes in their environment. Then respiration reactions coupled with photosynthesis made possible the complete and clean combustion of organic substances into carbon dioxide and water. Respiration released about four times more energy than did fermentation. Having much more energy available than they needed for survival, the living organisms benefited from an energy surplus. This is the capital that was invested in the immense enterprise of biological evolution.

In the evolution of societies the domestication of energy has taken place in three major stages, the third of which has scarcely begun. The first stage consisted of the long phase of human survival through the use of the earth's energy *income.* The second stage began about 150 years ago with the increased depletion of the energy *capital* of this planet. Finally, in our time there is the beginning of a progressive return to more efficient exploitation of the income of the ecosystem, associated with the planned use of its capital and the putting to work of nuclear energy.

The nomad hunter of prehistoric times was at the mercy of the great energy forces of the earth: fire, flood, storm, drought, and wild animals. Having barely enough energy for himself, he was unable to invest energy in the maintenance of even a rudimentary social organization. He could only collect the energy scattered throughout his environment and use it as he needed it, for he had no way of storing it.*

With the development of agriculture and metallurgy, mankind entered a phase of energy concentration. Grain and other foods were stored in baked earthenware pots. Pipes and canals channeled energy. Ovens concentrated heat, baked articles of clay, and smelted metals. People settled in fertile valleys and domesticated solar energy through the improvement of agricultural techniques. They gathered around large stores of food, establishing a defense system to protect themselves from the elements, wild animals, and other tribes. The presence of food in reserve freed some men from seasonal constraints, thereby allowing time for the production of crafts and for invention.

The concentration of men and energy led inevitably to the control of men and the control of energy. The exploitation of biological energy brought rule and servitude—of one man by another man. Slaves, galley slaves, and serfs were inexpensive machines that were easy to control.

Harnessing the physical energy of the elements led to the expansion of navigation by sail, the construction of canals, dikes, and dams, and the operation of wind and water mills. The improved efficiency of domestic animals, tools, and machines facilitated the processes of extracting and storing energy. Consumption increased and the rhythm of evolution quickened.

The transition to the second stage of the domestication of energy coincided with the discovery and use of the mineral resources coal and oil—the earth's capital. The subsequent period of the explosive exploitation of resources has been but a few moments in geological time.

Social organization continued, becoming more complex in cities. Coal, steam, and machines were in the ascendence as work became specialized and factories were built; this was the time of railroads and transatlantic ships. Industrial expansion at the end of the nineteenth century and the beginning of the twentieth saw the birth of capitalism, the establishment of a working class, and the new subservience of man through the work contract.

Oil is inexpensive energy; it made possible electricity, the automobile, and jet aircraft; it fostered the amazing explosion of industrial power, individual consumption, transportation, and communication. But it also led to the depletion of a precious capital, the abuse of the environment, and economic and political imbalances.

---

* See also the diagrams of the history of the economy, beginning on page 14.

In the face of the crisis, society's reaction is to call for the best available alternative source of energy: nuclear power. But this substitution is expensive in terms of capital, labor, and information—to say nothing of the new dangers it introduces. Caught between two great stages of human development, nuclear power may require even greater amounts of accumulated capital—money, goods, and technology—than the actual use of the earth's resources, even if fuel consumption were very low.

This short history of energy demonstrates that human societies have not escaped its implacable laws. The more complex a society, the more it will need significant quantities of energy to maintain itself. In every system in nature, organization continues until the energy cost of an increase in complexity is equal to the total energy budget available to the system. When this budget is exceeded, or when the sources of energy are exhausted, the systems become disorganized and disappear.

The same conditions apply to social systems. In a complex organization each individual is bound to the others by a complicated network of interdependent functions that involves exchanges of energy, materials, and work. Such an organization must divert to its own use a part of the energy budget that should have been distributed to each individual. In modern societies almost half of the energy received by individuals—in the form of wages, income, manufactured goods, and food—must be returned to the "organization" (the government) in the form of taxes in order to ensure the survival of the social system.

2. THE GREAT LAWS OF ENERGY

The principal energy laws that govern every organization are derived from two famous laws of thermodynamics. The second law, known as Carnot's principle, is controlled by the concept of entropy.

Entropy and the Science of Heat

Today the word *entropy* is as much a part of the language of the physical sciences as it is of the human sciences. Unfortunately, physicists, engineers, and sociologists use indiscriminately a number of terms that they take to be synonymous with entropy, such as disorder, probability, noise, random mixture, heat; or they use terms they consider synonymous with antientropy, such as information, neguentropy, complexity, organization, order, improbability.

There are at least three ways of defining entropy: in terms of thermodynamics (the science of heat), where the names of Mayer, Joule, Carnot, and Clausius (1865) are important; in terms of statistical theory, which fosters the equivalence of entropy and disorder—as a result of the work of Maxwell, Gibbs, and Boltzmann (1875); and in terms of information theory, which demonstrates the equivalence of neguentropy (the opposite of entropy) and information—as a result of the work of Szilard, Gabor, Rothstein, and Brillouin (1940–1950).*

The two principal laws of thermodynamics apply only to closed systems, that is, entities with which there can be no exchange of energy, information, or material. The universe in its totality might be considered a closed system of this type; this would allow the two laws to be applied to it.

The first law of thermodynamics says that the total *quantity* of energy in the universe remains constant. This is the principle of the conservation of energy.

The second law of thermodynamics states that the *quality* of this energy is degraded irreversibly. This is the principle of the degradation of energy.

The first principle establishes the equivalence of the different forms of energy (radiant, chemical, physical, electrical, and thermal), the possibility of transformation from one form to another, and the laws that govern these transformations. This first principle considers heat and energy as two magnitudes of the same physical nature (Fig. 68).

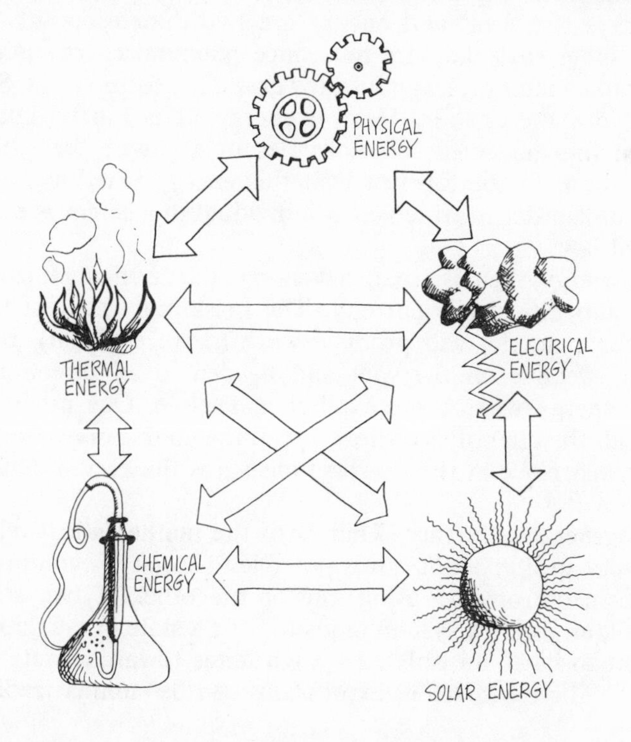

Fig. 68

* I shall consider here only the thermodynamic and statistical aspects of entropy. The relationship between information and entropy is treated in the following chapter, on information.

About 1850 the studies of Lord Kelvin, Carnot, and Clausius of the exchanges of energy in thermal machines revealed that there is a hierarchy among the various forms of energy and an imbalance in their transformations. This hierarchy and this imbalance are the basis of the formulation of the second principle.

In fact physical, chemical, and electrical energy can be completely changed into heat. But the reverse (heat into physical energy, for example) cannot be fully accomplished without outside help or without an inevitable loss of energy in the form of irretrievable heat. This does not mean that the energy is destroyed; it means that it becomes unavailable for producing work. The irreversible increase of this nondisposable energy in the universe is measured by the abstract dimension that Clausius in 1865 called entropy (from the Greek *entrope,* change).

The concept of entropy is particularly abstract and by the same token difficult to present. Yet some scientists consider it intuitively; they need only refer mentally to actual states such as disorder, waste, and the loss of time or information. But how can degraded energy, or its hierarchy, or the process of degradation be truly represented?

There seems to be a contradiction between the first and second principles. One says that heat and energy are two dimensions of the same nature; the other says they are not, since potential energy is degraded irreversibly to an inferior, less noble, lower-quality form—heat. Statistical theory provides the answer. Heat *is* energy; it is kinetic energy that results from the movement of molecules in a gas or the vibration of atoms in a solid. In the form of heat this energy is reduced to a state of maximum disorder in which each individual movement is neutralized by statistical laws.

Potential energy, then, is organized energy; heat is disorganized energy. And maximum disorder is entropy. The mass movement of molecules (in a gas, for example) will produce work (drive a piston). But where motion is ineffective on the spot and headed in all directions at the same time, energy will be present but ineffective. One might say that the sum of all the quantities of heat lost in the course of all the activities that have taken place in the universe measures the accumulation of entropy.

One can generalize further. Thanks to the mathematical relation between *disorder* and *probability,* it is possible to speak of evolution toward an increase in entropy by using one or the other of two statements: "left to itself, an isolated system tends toward a state of maximum disorder" or "left to itself, an isolated system tends toward a state of higher probability." These equivalent expressions can be summarized:

Potential energy ⟶ entropy
Ordered energy ⟶ disorganized energy (heat)
High-quality energy ⟶ heat (low-grade energy)
Order ⟶ disorder
Improbability ⟶ probability

The concepts of entropy and irreversibility, derived from the second principle, have had a tremendous impact on our view of the universe. In breaking the vicious circle of repetitiveness in which the ancients were trapped, and in being confronted with biological evolution generating order and organization, the concept of entropy indirectly opens the way to a philosophy of progress and development. At the same time it introduces the complementarity between the "two great drifts of the universe" described in the works of Bergson and Teilhard de Chardin.

The image of the inexorable death of the universe, as suggested by the second principle, has profoundly influenced our philosophy, our ethics, our vision of the world, and even our art. The thought that by the very nature of entropy the ultimate and only possible future for man is annihilation has infiltrated our culture like a paralysis. This consideration led Léon Brillouin to ask, "How is it possible to understand life when the entire world is ordered by a law such as the second principle of thermodynamics, which points to death and annihilation?"

Energy and Power

There can be no production of work without a previous concentration of energy or the existence of some reservoir of potential energy (such as the sun, the gasoline in a car, a hydroelectric dam, a storage battery, a steam boiler). This energy must then flow from the reservoir to a sink in which it is degraded and dispersed in entropy. Carnot's laws show that this loss of quality of potential energy is necessary to the functioning of any engine. The higher the drop in potential, the greater is the quantity of work produced. In a thermal engine this drop happens between the boiler (the hot source) and the condenser (the cold sink).

The law of potentials says that the flows of heat, electricity, or liquids that leave the reservoirs are a function of the size of the stored quantities. For example, the intensity of the current in an electrical circuit depends on the difference of potential between the generator and the resistance of the circuit. The flow of income from invested capital is proportionate to the total value of the capital.

To produce work from energy it is necessary to transform the energy from potential to actual. The quantity of useful energy released per unit of time is measured in units of power. The concept of power is therefore very general. We speak of the power of an electric generator, of a locomotive, of the sun's rays; we speak of the power of a country, an army, an economy, a political group. To release power requires a paradoxically small amount of energy in the form of operating energy, or information. The capacity to release large quantities of energy through the amplifying nature of information is commonly called the "power" of an individual. Thus human power controls physical power through information.

The amplification of information modifies the balance of power. To be applicable, the capacity for making decisions requires a tilting of the balance of power. For this reason any assembly, board of directors, or jury must have a majority, if only of one person. For the same reason the takeover of an organization often requires the gaining control first of the means for releasing power.

In 1922 A. J. Lotka proposed the interesting "law of maximum energy," which he applied to biological evolution. The law said that one of the factors that seem to have the most importance in the survival of an organism is the production of a large quantity of energy. This energy is used in maintaining the structure, in reproduction, and in growth. The creation of maximum power thus appears to be a *condition for survival* in the struggle for life. This law is also valid for human organizations.

Power and growth must then be the two principal factors of self-selection of a system. But power is released only at the expense of a significant waste of energy. This waste results from the lowering of efficiency in metabolic processes. And this efficiency is maintained at a remarkably constant level, even for open systems of a very different nature.

In all open systems the transfer of energy takes place by means of coupling processes: the input of one system is the output of another. The observation of very many reactions shows that the coupling process always develops an "optimum yield" that corresponds to maximum power. This yield is in the neighborhood of 50 percent of *ideal efficiency.*

Consider an example. The dropping of a weight produces enough power to raise another weight, to which the first is linked by a rope and pulley. It is a coupled process: the potential energy of the higher weight, transformed into kinetic energy by its fall, causes the storage (in the medium of the other weight) of a supply of potential energy. This energy in turn can produce new work (Fig. 69).

Obviously two equal weights will lead to no movement, no work, no storage of energy. However, if this ideal system should be set in motion, the efficiency of the coupling process would reach 100 percent. In fact the fall of the first weight, with the help of the second, will store an *exactly equal amount of energy.*

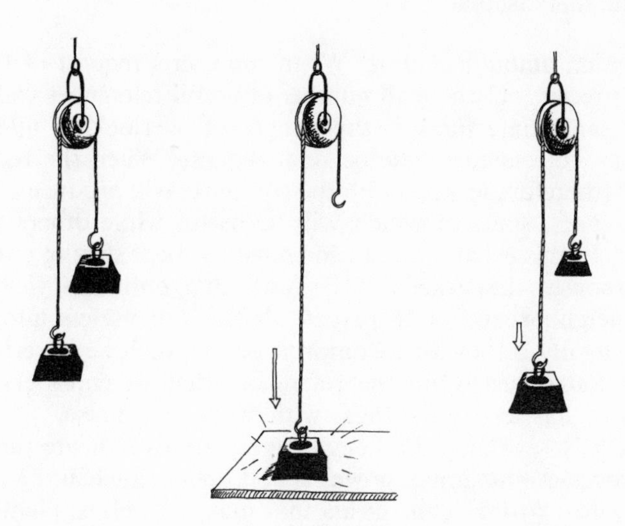

Fig. 69

Consider another extreme case. The right-hand weight is now equal to zero and the left-hand weight is released. There is a rapid fall and an impact on the ground. All the kinetic energy is lost in heat and thus no energy is stored.

These two extreme cases demonstrate that the only arrangement that will allow the simultaneous release and storage of energy *in a minimum of time* is that in which the weight on the right is equal to half the weight on the left. There is of course an energy loss in the form of heat at the moment of impact, but the maximum load is raised in the minimum amount of time. If the lighter weight were heavier, the process would be extremely slow. If it were lighter, too much energy would be lost in heat at the moment of impact and the yield would be still less.

The generalization of this principle to an entire category of irreversible processes that take place in open systems was proposed during the 1950s by physicists of the school of "thermodynamics of irreversible processes." Their work showed that each of the magnitudes such as voltage, temperature gradient, gravity, and concentration can be considered a thermodynamic force (or potential). Coupled to each force is a flow whose speed is proportional to the force that determines it. In the examples above, this flow would be an electric current, heat, the speed of a body in motion, or a flow of molecules. Once again one discovers the state and flow variables and their controls, encountered in the preceding chapters.

An interesting approach to the law of optimum yield can be made on the ground of information. The power of a computerized data bank for storage and retrieval of bibliographical references can be expressed as the maximum amount of useful information that can be obtained in a minimum amount of time. When the user's request of the data bank is very precise, only a small number of useful references will be obtained. At the same time there is the danger of overlooking many references that may be related to the original request. When the request is more general (therefore less precise), the computer will produce a large number of references, some of which will be useful while others will be of no interest. In this event a lot of time must be spent sorting out the computer's responses. Experience has shown that optimum yield is obtained in the neighborhood of 50 percent "noise," or useless information. The user generally settles for a compromise in which he is certain of having rapidly found nearly all the references that he considers useful, even though he has to pay for them with 50 percent noise.

Generally speaking, mathematical equations indicate that, in all coupled processes, maximum power is best obtained when the ratio of forces is equivalent to 1:2. This means that man (as well as plants or animals) *prefers to sacrifice yield to power.* This is readily seen in the consumption of energy in society. This very simple and very general law, which applies equally to physical, biological, and social systems, is expressed by Figure 70.*

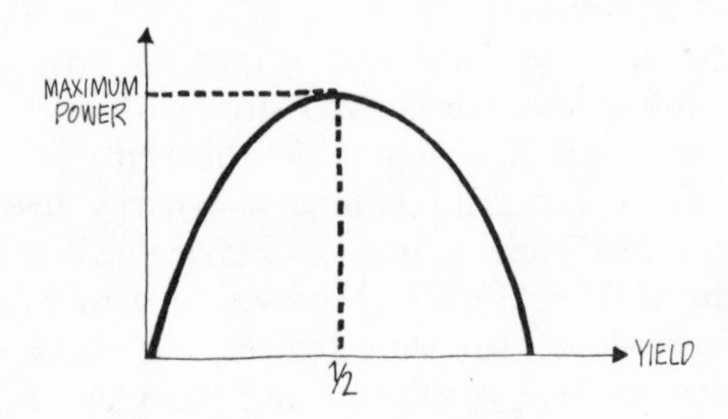

Fig. 70

Another law that applies to a large number of systems is the law of diminishing returns that is so well known to economists. When a global result is obtained by the multiplication of several factors, the growth of *only one of them* need be limited to cause the global result to stretch also toward an unreachable or asymptotic limit. The mathematical function that illustrates this law is a hyperbole. In spite of the very important increase in the quantities represented on the horizontal axis, the yield represented on the vertical axis no longer increases (Fig. 71).

---

* The identification of flow and force in social and ecological systems is quite recent. Ecologists like Howard Odum are responsible for having identified it in a large number of processes.

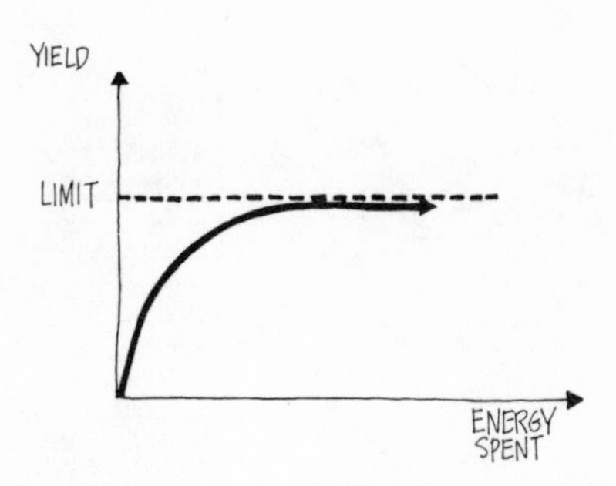

Fig. 71

One encounters such law in biology, in the saturation of active sites of enzymes; in agriculture, where in spite of the massive injection of fossil fuels the energy yield of agricultural processes (in terms of calories consumed by man) reaches a limit; in accounting, where the efforts used to obtain two figures after the decimal point are far superior to the actual utility of such precision in the accounts; and in navigation, where after a certain speed has been reached no effort of the crew or addition of sail will have more than a marginal effect on increasing the speed.

The lesson of the law of diminishing returns is severe: the limit of return has long since been attained—whether or not the fact has been realized—in numerous organizations, business firms, and work teams. Nevertheless, continuing to try to improve returns, man expends prodigious amounts of ingenuity, large quantities of energy, and important human and material resources—while the limiting factor remains completely unnoticed.

3. METABOLISM AND WASTE IN THE SOCIAL ORGANISM

Like every living organism, human society transforms, stores, distributes, and wastes energy in order to survive, produce work, and evolve. The circulation of energy in its structures and the transformations that take place there are its metabolism.

Metabolism includes all "biological machinery," human and animal, and all mechanical and electronic equipment that men use in their social activities. The biological machinery depends on food, the mechanical machinery on oil and electricity—more generally, on fossil fuels. The biological machinery and the world population of mechanical and electronic machines all transform energy into useful work, thereby allowing the maintenance and development of the social organization (Fig. 72).

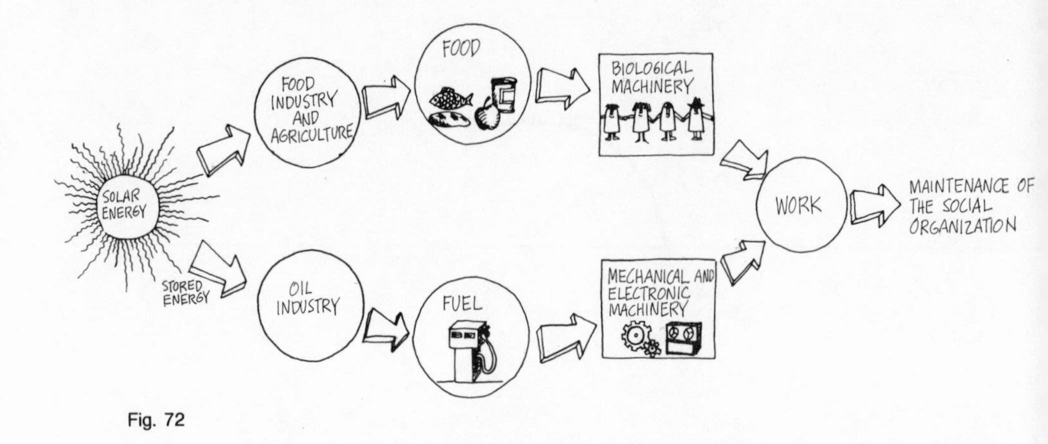

Fig. 72

The metabolism of a man walking at a normal pace consumes 200 watts. His minimum energy requirement is about 1,320 kilocalories per day, or 2,500 kilocalories per day for moderate activity.* With the help of fire, prehistoric man used 4,000 kilocalories, or twice the amount of energy needed for his metabolism. In a primitive agricultural society man consumed about 12,000 kilocalories; with the beginning of the industrial era, he used about 70,000 kilocalories. Today the average American uses 230,000 kilocalories per day. A well-known statistic, one that it is helpful to recall, is that America, with only 6 percent of the world's population, consumes 30 percent of the world's energy, or 20×10^{15} kilocalories.

The biological machinery of a country like France is made up of 50 million individuals (not counting the animal population). It produces 4.3×10^{10} hours of effective work annually, or the equivalent of about five million years of work.† It consumes in food energy alone 45×10^{12} kilocalories of food per year. It produces 35,000 tons of waste per day and 12 million tons of garbage per year. If you also count industrial and commercial wastes, 25 million tons of waste are produced each year. Annual energy consumption per capita reached 32 million kilocalories in 1973, or a total annual consumption of 1.6×10^{15} kilocalories for all France.

Other than food calories, all energy spent by society is consumed by the machinery used by men. World consumption of energy was 58 ×

* The kilocalorie is one thousand times greater than the "small calorie," or the quantity of energy needed to raise by one degree a cubic centimeter of water at 15° Centigrade.

† There are about 20,900,000 active persons working an average of 173 hours per month.

10^{15} kilocalories in 1974, and it will probably reach, with a rate of growth of about 4 percent per year, 100×10^{15} kilocalories just before the year 2000.

Thermal energy, released mainly through the combustion of oil and its derivatives, turns the motors of machinery, automobiles, and electric generators. All these machines can be grouped in four large categories: transportation, industry, commerce, and domestic use.

The aggregate flow of the circulation of energy in the social organism follows the law of optimum yield found in all open systems. The total energy yield of social systems seems to stabilize at about 50 percent, or an amount corresponding to the production of maximum power required by the intensity of their metabolism. In all large, developed countries, the overall return seems to be 50 percent. Figure 73 below shows the use of energy in American society.

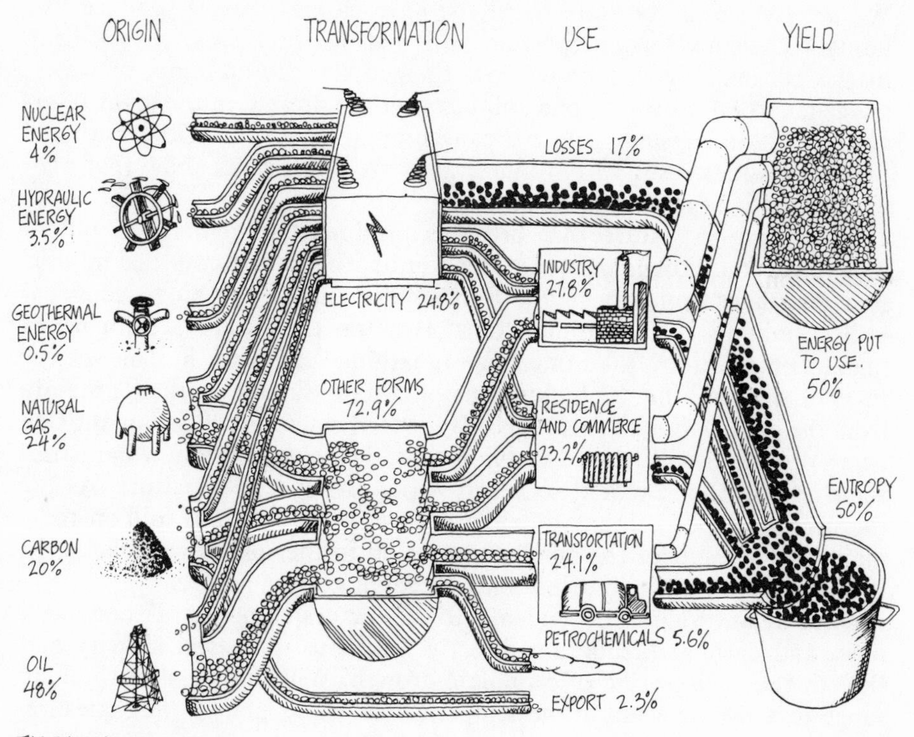

Fig. 73

All forms of metabolism produce wastes and entropy. But with the acceleration in the consumption of energy and the consequent acceleration in the intensity of metabolism of the social system, the action of man on nature assumes alarming proportions. Three major energy crises have arisen in our industrial civilization as a result of man's activity: the energy (and raw materials) crisis, the food crisis, and the environmental crisis.

A truly systemic approach to the energy problems must be worldwide in scope and must hold for the long term. Thus I prefer to consider in broad terms the possible long-term effects of man's energy-related activities on the climate of our planet rather than describe the effects of specific pollutants.*

The total quantity of gaseous, solid, and liquid wastes produced by the social organization's metabolism now reaches proportions that are about equal to the total quantity of elements recycled by the ecosystem. We are long past the time when the aggregate of wastes produced by humanity seemed, in comparison with natural processes, only a drop in the ocean. Now we know how to measure the amounts of water, oxygen, carbon, nitrogen, and sulphur that are present in the great reservoirs of the ecosystem, and we can compare these quantities with the production that results from man's activity. The results show that man is in direct competition with nature.

Everything in nature that breathes produces altogether 720 billion tons of carbon dioxide per year. In addition, the total amount of sulphur that circulates annually in the atmosphere and in the biogeochemical cycles amounts to 500 million tons. How do these figures compare to man's contribution? Like one huge breathing organism, human society in 1977 was exhaling 22 billion tons of carbon dioxide annually, mainly from the combustion of fossil fuels. By 1980 the amount of carbon dioxide being produced will reach 26 billion tons. The contribution of industrial societies will then amount to nearly 4 percent of that of nature. As for the production of sulphur dioxide, it rose in 1975 to 760 million tons, mainly as a result of the combustion of oil in thermal power plants. In the year 2000 man's contribution will equal that of nature.

Once again the interdependence of the factors eludes us. Excess heat, dust, and carbon dioxide are directly linked to industrial activity and thus to the acceleration of economic growth. Will they lead in the long range to a rise or a fall in the temperatures of our planet? The question

---

* Specific kinds of pollution are those caused by carbon monoxide (CO), sulphur dioxide (SO_2), and nitrogen oxide (NO); one might also include water pollution, solid wastes, radioactive wastes, and thermal pollution of the rivers. Norms and standards established by governments and international organizations generally apply to these types of pollutants.

is often discussed; what effect would it have? the cooling that results from atmospheric dusts or the "greenhouse effect" of carbon dioxide? These questions are difficult to answer; the earth also readjusts its equilibrium, though not always to our satisfaction.

All of the world's energy ends in the form of heat. First it is stored in the biosphere and in water, then it is dissipated in the atmosphere and radiated into space. The total amount of heat released by human society can easily be measured because it is a function of the consumption of energy, and that is a well-known figure. Climatologists estimate that there could be significant changes in the world's climate if the heat released by man were to reach one percent of the amount of energy that comes to us from the sun. Yet the heat produced by the fourteen northeastern states, which consume 40 percent of all energy in the United States, amounts to 1.2 percent of the total energy received from the sun over the same area. The figure will rise to 5 percent by 2000—a figure already reached in Manhattan alone.

Local climatic modifications have already been observed. They can be seen in the cloud formations downwind from large power plants. A power station producing 20,000 megawatts can cause storms and heavy rains, as was demonstrated by studies in St. Louis in 1973. The modification of the microclimate of large cities is also visible (see p. 32). It would appear that city dwellers are condemned to perpetually gray and humid weather in winter and unpleasant weather in summer, all because of thunderstorms. And the climate of the entire world may suffer the same fate, for the problems created by the evacuation of heat in the atmosphere are becoming serious. The use of energy grows at an exponential rate: nearly 4 percent per year in the United States and 6 percent per year in the rest of the world. At this rate the limit of one percent of the sun's energy will be reached over the whole world in 130 years.

To the effects of excess heat we must add the greenhouse effect of carbon dioxide. Shortwave radiation coming from the sun easily penetrates the layer of carbon dioxide that surrounds the earth. But infrared rays reflected by the earth's surface cannot penetrate this layer; being trapped, they contribute to the warming of the atmosphere. The amount of carbon dioxide in the atmosphere, as a result of man's activities, increases by 0.2 percent per year. Presumably the greenhouse effect could raise the average temperature of the earth. But this is the reverse of the general cooling we have seen since 1940.

Perhaps the key to this paradoxical situation is to be found in the factors that increase the reflection of solar radiation by the earth. This reflection is called *albedo.* The regulatory role of albedo is a determining factor in the thermal equilibrium of our planet; because of it the earth maintains its temperature in a stationary state. The difference in temperature between the equator and the poles remains almost constant.

We learned several years ago, by means of photographs taken by meteorological satellites, that the surface area of the ice in the northern hemisphere exceeded by 12 percent the areas measured in the preceding years. Ice formed earlier and melted later. The differences were very marked beginning in 1972 and 1973, a period when, according to the *Bulletin of the World Meteorological Organization,* an abnormal climate was observed.

According to Reid Bryson, director of the Institute of Environmental Studies at the University of Wisconsin, the cooling of the earth may be due to the increase in the quantity of dust particles and aerosols in suspension in the atmosphere. These dusts are dispersed in the atmosphere by natural causes (volcanos, desert winds, salt) and by human activity. Today their quantity totals 296 million tons (four million comes from volcanos). Of this total, about 15 million tons remains permanently in the upper atmosphere. Another two million tons could reduce the earth's temperature by 0.4° Centigrade. This is why the effect of dusts on the transparency of the atmosphere appears to be more important than the greenhouse effect of carbon dioxide.

One final factor comes into the picture. The increase in albedo has a more marked effect on the cooling at the poles than on the temperature of the tropics (because the sun's radiation travels laterally and penetrates a thicker layer of dust at the poles). The result is a much greater difference in temperature between the poles and the equator. The thermal machinery of the earth tries to equalize the difference, and this may well cause confusion and disturbance in the pattern of the prevailing winds.

In spite of rather troubling climatic conditions, one need not dramatize or despair. It is not yet certain that these disorders are the result only of the factors mentioned. Perhaps the world's temperature is passing through cyclical phases of which we are unaware. Then, although limited, the ecosystem is not static but is in dynamic equilibrium; its multiple stationary states can readjust themselves according to modifications resulting from human activities. Carbon dioxide, for example, whose concentration in the atmosphere continues to increase, is probably transformed little by little into carbonates and organic matter; that is, it is stored in the earth's sediments or in the wood of the forests.

4. ECONOMICS AND ECOLOGY

The illusion of continuous economic growth is nourished by the false notion that the economy is an isolated cyclical process that escapes the energy laws of the physical world and the increase of entropy. The opposition between the circular monetary flows and energy flows, which move in opposite directions, and the potential for creating money have probably helped to enrich and reinforce this illusion. Perhaps we should look further, to the collective subconscious, to the roots of that mad dream of humanity that tries to balance and even reverse through economic growth the natural aging process of social organizations. It is a pathetic struggle against death.

In the end one must pay. The bill for growth has just been given us, and it is enormous. Our natural resources are being exhausted; the environment is endangered; inequalities, far from being removed, are greater than ever.

In fact there is nothing cyclical in the economic process. As in all open systems, the circular movement in reality is an irreversible, one-directional process—that of the degradation of energy and the increase of entropy. It is not surprising that in the "classical" economy the relationship between economics and ecology—better still, between energy, ecology, economics, and entropy—has long gone unnoticed. And this is the question at hand, not that of the economy of the environment. Economics and ecology are like a series of interconnected reservoirs; to draw from one more quickly than it can be refilled means emptying it eventually.

The classical economy is defined in terms of the distribution of scarce commodities. The ultimate resource, whose scarcity conditions that of all the others, is *free energy.** The economy of biological and ecological systems is built entirely on the recognition of the importance of this ultimate good. This economy is based on the management of an energy capital and the judicious use of information in order to "organize" the energy into products that can be assimilated directly by the cell, the organism, or the different species of the ecosystem.

Should not the traditional economic act be enlarged and enriched along the lines of this fertile relationship between energy and information? In the new ecoenergetics viewpoint, the economy should deal with the management and equal distribution of an energy reserve and an energy flow, coupled with the "information" (transformation) of this energy into goods and services useful to society.

Universal Currency: The Kilocalorie

A systemic approach to the processes that link the economy and ecology must try to go beyond the already outdated concept of monetary value and complete it with the concept of energy cost, expressed in a universal unit of energy. This unit might be the kilocalorie; it would

* A definition of free energy is given in the note on pp. 116–117.

allow, at the level of the control and use of energy, a unification of biological, ecological, and socioeconomic systems.

The table below gives several estimates of magnitudes in terms of kilocalories.

Energy from the sun	3.7×10^{18} Kcal/day
Total world consumption, 1974	58×10^{15} Kcal
Food consumption in France	45×10^{12} Kcal
Energy needed to produce:	
aluminum (one ton)	50×10^{6} Kcal
paper (one ton)	10×10^{6} Kcal
steel (one ton)	7.5×10^{6} Kcal
cement (one ton)	2.2×10^{6} Kcal
petroleum (one ton)	1.3×10^{6} Kcal
Energy needed to produce a car (1.5 tons)	32×10^{6} Kcal
Energy needed to feed a man for 30 years (subsistence only)	32×10^{6} Kcal
Energy needs of a man	2,500 Kcal/day
Energy from one liter of gasoline	10,000 Kcal
Energy cost for one passenger on a transatlantic flight	6×10^{6} Kcal

One of the keys to ecoenergetics is the determination of the economic value of the kilocalorie. Is it possible to arrive at a conversion factor that links monetary value and energy cost? Even a rough estimate would be very useful. One might begin with graphs showing the relationship between gross national product and energy consumption per capita and draw from them an estimate of the dollar value of the kilocalorie according to the various economies.

In 1971 Howard Odum proposed an energy equivalent of 10,000 kilocalories (the energy of one liter of gasoline) per dollar. Today the equivalent would probably be no more than 5,000 to 7,000 kilocalories per dollar. Nevertheless such an approximation makes it possible to compare energy flow and money flow more efficiently. It also makes it possible to determine the cost of noneconomic goods such as living trees, water, and oxygen. The comparison of dollars and kilocalories also clarifies the concept of *energy-intensiveness,* the energy demand of certain industrial processes. This demand is stated in terms of the relationship between kilocalories consumed and value added.* An energy-intensive industry spends up to 50,000 kilocalories per dollar in value added.

The transition from economics to ecoenergetics is justified in many other ways. In ecology one finds the economic equivalent of the fee

---

* The value added of a product is the difference between the final value of the product and the value of all the materials used in its manufacture.

paid for finished work. This "payment" is based on an energy "price," and the "money" circulates in the form of materials useful to the community.

When a function is necessary to the maintenance of the structures of an ecosystem, the "reward" circuits, based on the mutual benefit of the species and the communities, are reinforced by feedback loops. Animals return to plants mineral substances (phosphates, nitrates, potassium) that are useful to plant growth. The "work" of animals (hunting, destroying, control of certain species, transfer of information) is the equivalent of a "service" paid for in food.

Regulation by kilocalories exists everywhere in the world of the producers and consumers of the ecosystem. Producers are stimulated when the flow of mineral substances that is returned to them is greater than the flow of food that they manufacture. Consumers are stimulated when the flow of food that they receive is greater than the flow of mineral substances that they return to the plants. But every balanced interdependence must depend on self-stimulating loops; in other words, the agents that participate in these loops must be rewarded. The stimulation of the transforming agents can be thought of as an "individual" motivation. Without reward or stimulation, an energy circuit dries up and disappears. Through the play of reinforcement loops and their interconnections, the ecosystem selects those species and individuals who contribute most effectively to the functioning and maintenance of the whole.

Finally, the ecoenergetics approach brings out the link between time and energy. The empirical law is simple: a gain in time must be paid for in energy. If we want to travel fast, we use a car or a jet; if we want to produce fast, we use assembly lines and automation. For the minutes saved we must compute the kilocalories spent. To save time, one increases the amount of energy subsidy fed into the social megamachine. This vicious circle of economic growth still believes that it liberates time. It saves time, but at what expense? in the face of what deadline?

Energy Analysis

The basic tool of ecoenergetics is *energy analysis.* This method will probably turn out to be one of the most fruitful in determining which solutions to apply to the crises we meet.

In order to find and put to work new sources of energy or to choose the most advantageous ways of saving energy, we must first be prepared to set up complete and detailed energy balance sheets. Today we can do this, thanks to new techniques of energy accountability that have come chiefly from chemical engineering, biology, and ecology. Together these techniques make up energy analysis.

The forerunners of energy analysis were Raymond L. Lindeman of

Yale, who in 1940 drew attention to the quantitative relationships that exist in nature between the various consumers of an ecosystem, and Howard Odum of the University of Florida. In 1957 Odum published a famous article, now a fundamental document of ecologists, in which he made a complete analysis of the energy flow (in kilocalories per square meter per day) that circulated within an ecosystem composed of the flora and fauna of a little river.

Other ecologists have applied these methods to the energy accounting of small communities dependent on hunting and fishing (Eskimo and African villages), and they have been able to connect energy factors to economic elements. But the true birth of energy analysis—and with it the beginnings of ecoenergetics—coincides with the reinterpretation of the results of economic analysis in terms of energy units.

The father of economic analysis is Vasilli Leontieff of Harvard, winner of the Nobel Prize in economics in 1972. As early as 1946 Leontieff had made an input-output matrix based on thirty sectors of the American economy. Input-output matrices are tables made up of a large number of entries which correspond to the different sectors of the economy. They follow and measure, from producers to consumers, the variations in supply and demand for raw materials, semifinished and finished products, and services. The results of economic analysis, expressed in monetary units, make it possible to develop the concept of value added.

Energy analysis is an offshoot of economic analysis but draws its inspiration from methods used in chemical engineering. It tries to estimate the energy cost of every industrial transformation that uses energy, raw materials, or work. Working backward, one retraces one by one the stages in the manufacture of a given product, constructing a tree whose branches become increasingly ramified. The amount of energy used at each step is measured, and in the end all the kilocalories expended are added up. The first applications of energy analysis were made in the automobile industry in 1972 and in food production in 1973.

The calculation of the energy costs in the manufacture of a car in the United States was made by Stephen Berry and his team from the chemistry department of the University of Chicago in 1972. To build an automobile weighing 1.5 tons requires an expenditure of 32 million kilocalories. Yet thermodynamic calculations show that in theory the necessary quantity is only 6 million kilocalories. The 26 million kilocalories of excess energy (80 percent of total consumption) is used only to save time. Thus the automobile industry consumes more free energy than it needs—in order to increase efficiency in production, to lower prices, to sell more cars, and thus to realize greater profits.*

* Unlike energy, free energy, or thermodynamic potential, introduces a quantitative expression of the desirability of a product. For example, the value of an iron ore is higher when the ore is rich in iron than when the ore is diluted a thousand times in dust. The difference between energy

The difference between economic analysis and energy analysis is a difference on the time scale. If economists were to determine the depletion of resources by extrapolating from longer and longer terms, their estimates would catch up with those of the thermodynamicists.

Economic estimates are based on the measure of energy needed for an infinitely slow and reversible process. Yet the economic system is an open system crossed by an irreversible flow of energy. Moreover, as the law of optimum yield indicates, we prefer to sacrifice yield to power—which causes us to waste, on the average, about 50 percent of the available energy in order to achieve more rapid change. Supplementary energy, which makes possible the transfer from thermodynamics to economics and which measures at the same time the "value" we assign to material things, is the *energy subsidy.* This subsidy says in energy terms that the cost is directly connected to the intensity or speed of the transformation. The aggregate energy subsidy is related to the speed of the metabolism of the social organism and therefore to the rhythm of its growth. It is well known that countries with the highest growth rate and the highest GNP are the largest consumers of energy.

The increase in free energy at each stage in the transformation of a product during its manufacture is the physical equivalent of the economic concept of value added. Energy reaches a maximum at the moment of purchase by the consumer, then declines more or less rapidly. But the consumer does not "consume" the product; he throws it back into the environment as soon as he decides it is useless. Thus some disposals retain a high level of free energy. The real waste is in the nonutilization of the free energy that remains in the discarded products.

Energy Analysis and Food Production

One of the most revealing applications of energy analysis has been in the entire area of food production. The production and distribution of food are among the most important functions of the social organism. This importance is indicated at the economic level by that part of the family budget allotted to food. In France it fell from 49 percent in 1950 to 27 percent in 1973; in the United States it accounts for less than 22 percent of the household budget. A money flow that represents about one-fourth of the total budget of the consumer sector must be balanced by an energy flow of at least equal amount.

Energy analysis shows that the total amount of energy spent on food production represents about 15 percent of the total energy budget of the United States and 22 percent of the total electricity budget. This

---

and thermodynamic potential brings in entropy, that is, entropy multiplied by the absolute temperature of the transformation.

free energy = energy − entropy × absolute temperature

energy is used in farm operations, processing industries, transportation, markets and stores, and at home in refrigerators, freezers, kitchen stoves, and ovens. In 1973 Americans used six times more energy to feed themselves than was necessary for human metabolism. The rate of increase in the amount of energy needed for food production is higher than the population growth rate in the United States. All this serves to promote energy analysis and to pose long-term questions about the harmony of the formidable machinery of food production that supports the populations of the developed countries.

If one considers the different ecological chains and cycles from the transformation of solar energy in green plants to the lump of sugar in our coffee, the fat in a piece of meat, or the bread for breakfast, one sees that the growth and maintenance of agricultural yields required by demographic pressure and the rise in the standard of living have been made possible, just in the last fifty years, only by massive injections of fossil fuels into the agricultural processes. Considering all these processes together as one machine, it comes down to using more natural "solar" calories and ever more "fossil" calories as input in the hope of recovering in the output calories that can be consumed by the living organism.

Quite simply, the question is this: are we not spending, in our industrialized societies, more calories in input than we recuperate in output? In other words, is the input of fossil fuels greater than the output of calories in agricultural production? Above all, doesn't the ratio of calorie output to calorie input tend to diminish in disturbing proportions? If the answer to these questions is affirmative, then we can expect—as our industrialized countries are already aware—a shortage of calories perhaps less dramatic than that in the underdeveloped countries but leading nevertheless to uncontrollable increases in the cost of food products.

Applied to agriculture, energy analysis shows how energy from fossil fuels substitutes for the energy otherwise provided by the work of men, domestic animals, and the natural elements. Fossil energy also substitutes for natural fertilizers, but the manufacture of nitrates, phosphates, and potassium fertilizers requires a higher expenditure of energy. Pumps used to irrigate crops, formerly operated by animal or wind energy, have been replaced by electric pumps and diesel engines. Instead of sunlight, fuel and electricity are used to dry fodder; they are also used to light and air-condition special barns devoted to the intensive raising of livestock.

The most significant results of energy analysis come from measuring the total energy input necessary to produce a given food product. David

Pimentel and his group from the College of Agriculture and Biological Sciences of the State of New York sought to determine, over a period of twenty-five years (1945–1970), the increase in the energy subsidy needed to grow one acre (0.4 hectare) of corn in the United States. They included in their energy analysis the labor of agricultural workers (kilocalories consumed per day); the energy cost of the manufacture of farm machines; the gasoline consumed; the energy cost of the production of fertilizers, insecticides, herbicides, and seeds; and the electricity or fuel used in the irrigation, drying, and transportation of the corn.

Their analysis revealed that in 1970 it required 2.9 million kilocalories per acre (the energy equivalent of 750 liters of gasoline per hectare) to produce the 8.16 million kilocalories contained in the harvested corn. Thus the energy yield was the equivalent of 2.82 calories per calorie invested. But in 1945 it had been 3.2 calories per calorie invested. Between 1945 and 1970, then, the energy yield from growing corn decreased by 24 percent, while the actual yield, in tons of corn per hectare, increased regularly.

Energy analysis was extended further by John S. Steinhart of the University of Wisconsin, who applied it to the entire agricultural and food-producing system of the United States between 1940 and 1970. The results showed an increase in the gap between the energy needed for food production and the energy equivalent of the food needs of the American people during those years. The gap widens because the energy consumption connecting agriculture and the increase in the standard of living involves a growing use of canned, frozen, and prepared foods whose preparation or storage requires significant quantities of energy. This increase also reflects a higher consumption of food outside the home—chiefly at places of work—and the fact that beef has a low rate of efficiency in transforming "solar" calories into "food" calories.

One of the most disturbing results is that we are approaching little by little the theoretical limits of the agricultural yield—an excellent illustration of the law of diminishing returns. The slope of the curve that traces the amount of energy needed to produce one food calorie, in the course of the history of United States agriculture, at no point becomes less steep, which confirms the fall in efficiency in the entire production process.

Such an evolution should be compared to that of the underdeveloped countries, or "primitive" cultures, where an investment of one calorie brings from five to fifty food calories in return. In our developed countries it takes from five to ten calories in fossil fuels to produce one food calorie.

The Competition Between Energy and Work

Energy analysis has also been applied to the problems of pollution caused by solid wastes, in attempts to determine the more advantageous of two solutions: to collect old papers and cartons for recycling or to burn them and use the energy for heating buildings.

There are interesting perspectives in the study of the consequences of substituting energy for human labor, the creation and elimination of employment, and the changeover from an energy-intensive manufacturing process to one that is labor-intensive.

There is a very close relationship between energy, labor, and production capital. Economists have long known that energy and labor vary in inverse direction. They compete for the same share of production capital. Because energy has the capacity to provide work, energy and labor are substitutes for each other.

Some examples will illustrate these relationships and their significance for energy analysis. The five industries that are the largest consumers of energy are aluminum, paper, steel, concrete, and petrochemicals. Together they consume 40 percent of all energy used in the industrial sector in the United States, yet they employ only 25 percent of the total labor force. Production capital is very important in these industries; heavy equipment, complex machinery, and automation all require high capital investment. But wherever energy is substituted for human labor, a more significant amount of dividends flows toward the stockholders. When energy prices are low, the cost of production capital is high; when prices increase, production capital shrinks.

There is a consequence of the competition between energy and labor. If energy prices continue to rise, the long-term tendency will show—paradoxically—increases in employment and in wages. All things being equal, the share of national income that goes to reward labor will increase at the expense of capital income. From 1947 to 1971 the low cost of energy made it possible to replace manual labor with energy whenever it was feasible. The consequence, inevitably, was an impoverishment of the quality of work in the expansion of assembly lines and bureaucracy.

We can also use the economic input-output matrices, retranslating them into energy values in order to measure the total quantities of energy and labor (as goods and services) needed to reach a given level of production. Bruce Hannon of the University of Illinois did this in 1974.

To produce an additional $100,000 of aluminum, for example, 9.5×10^9 kilocalories and five persons would be needed. To produce the same supplementary value in tobacco (an industry that uses little energy) would require only 1.2×10^9 kilocalories and the creation of thirty-two new jobs. Thus a transfer of $100,000 in consumer demand from aluminum to tobacco would reduce the consumption of energy by 33 percent and create twenty-seven new jobs.

Highway construction is one of the major consumers of energy because of its use of petroleum-based asphalt and concrete, whose production requires large quantities of energy. A $5 billion highway construction program consumes 55.4×10^{12} kilocalories and provides jobs, directly or indirectly, for 256,000 people. The same amount invested in a railway system would consume 20.1×10^{12} kilocalories and provide employment for 264,000 individuals—a gain of 8,000 employees over the highway construction program. And the same amount spent on a vast program of public health would certainly use energy, but it would make possible the creation of 423,000 jobs (167,000 more than the highway program, 159,000 more than the railway program).

Energy analysis will help in a positive way to make it easier to choose the most appropriate and advantageous means for resolving some of the problems created by the energy crisis, the food crisis, and the environmental crisis. Energy analysis will enable us to answer—with the figures to support our answers—questions of this type: Is it more costly to implement new sources of energy or to improve efficiency in the production of aluminum?

5. BIRTH OF THE BIOINDUSTRY

There are three principal ways to reduce our consumption of fossil fuels and raw materials: the implementation of new energy sources, the recycling of materials, and energy conservation. In the long term these will involve a transition to products whose manufacture consumes less energy, a greater emphasis in the economy on the services sector, and the implementation of "soft" technologies. These ways are well known; they are mentioned again only to emphasize the extent of the transformations through which society is passing.

The new energy sources, of which there is so much talk, are chiefly nuclear energy (fission and fusion), solar energy, and geothermal energy. Energy from the combustion of organic products, wind energy, or waterfalls can be considered derivative forms of solar energy.

The transformation of solar energy into heat is made possible by heating and air-conditioning systems. It can be transformed directly into electricity by photoelectric cells. Indirectly this energy can be released by the burning of organic waste, by the production of liquid fuels through chemical decomposition and of gases (methane) through bioconversions of manures by bacterial fermentation.

The recycling of wastes must be incorporated in a much more general process of the recovery of discarded materials—the equivalent in society of the recycling achieved in the ecosystem by the decomposers (see p. 8). Recovery includes the reuse of objects and the recycling of materials in production. Discarded materials can be grouped in two categories, wastes and debris (materials that have no use in a given economic context). In the long term the only valid means for rebuilding the great natural recycling loop is to involve the population in sorting out materials when they are discarded. For a small expenditure of energy and with careful use of information, each individual can reduce the entropy in a heap of discarded items. Machines can do this only at prohibitive cost.

Energy conservation is accomplished mainly through the retrieval of heat, through thermal isolation, and through the replacement of energy-intensive industrial processes and energy-intensive transportation systems by more economical means. At least 25 percent of the world's energy could be saved by observing a few basic rules of energy conservation.

No matter how ingenious the solutions or how effective the discipline of the population when confronted with problems of waste, the real long-term solutions will come only with a radical remodeling of our way of living in society—living differently and living in cooperation with nature. This new perspective involves the bioindustry and ecoengineering.

The coming revolution in agriculture and in the food and chemical industries will be of a *biotechnological* nature. It will give birth to a bioindustry that will bring new solutions, based on soft technologies, to the energy crisis and the degradation of the environment.

After the advent of agriculture about ten thousand years ago, the first agricultural revolution took place in the seventeenth century. It was marked by the techniques of crop rotation and the selection and hybridization of seeds. The second revolution came in the middle of the twentieth century with the mechanization of agriculture. The third revolution, now in preparation, will be based on biological engineering, new methods of energy conservation, and the controlled use of natural cycles.

Better insight into the development of microorganisms has already made it possible to produce proteins by growing yeast on hydrocarbons such as methanol and methane. The use of insect hormones to sterilize male insects assures the control of populations of insect pests—and this for a fraction of the energy cost required to produce pesticides.

But we must go further. Biological information collected over the last thirty years makes greater development possible. The agricultural and industrial revolution of the end of the twentieth century will depend on techniques that have hardly left the research stage: genetic engineering, enzyme engineering, bacterial engineering. There will be synthetic molecules performing the activity of natural enzymes, fermentation reactions controlled by computers, control and use of the basic reactions of photosynthesis, and (why not?) abiotic syntheses like those of the primeval earth, when the first molecules of life appeared.

This revolution will see the appearance of a new form of slavery: the domestication of microbes, those docile and indefatigable workers. In order to replace man or some of his machines in numerous tasks, we can consider two directions, each corresponding to a form of slavery: the sophisticated electronic system of industrial robots (tremendous energy consumers) or biology and the slavery of the myriads of microbes that populate the biosphere. These two routes are already being traveled today, and it appears that the bioindustry and the domestication of microbes will be recognized as even more spectacular developments than the use of industrial robots with their great appetite for energy.

This revolution will free agriculture and the food industry from the vicious circle in which they are now confined by the exhaustion of energy resources, the decrease in the calorie yield of the agricultural machine, and the concomitant increase in the price of calories consumed by men.

Four sectors will probably dominate the bioindustry in the years to come: the production of chemical products by microbes; the domestication of enzymes; the electronic control of fermentation reactions and bioconversions in general (capable, for instance, of generating energy); and the control of the basic reactions of photosynthesis.

New Jobs for Microbes

There are not only pathogenic microbes, there are useful ones as well. We have long been putting them to work to ferment wine and beer, to make bread dough rise, and to produce yogurt and cheese. Under the general title of decomposers, these microorganisms are the recycling agents of the ecosystem.

Today microorganisms are finding new employment in industry. They are used as miniature factories in the manufacture of dozens of commercial products such as amino acids, enzymes, solvants, insecticides, and antibiotics. The energy crisis and the food crisis are accelerating this mobilization of useful microbes.

Biological processes in nature are controlled by catalysts of amazing efficiency—the enzymes. Because of them the reactions of life occur at room temperature and in mild conditions, in complete opposition to the energy-intensive processes of the chemical and food industries. Moreover, the by-products of the metabolism of microbes are either useful substances or harmless molecules like carbon dioxide and water.

The secret of microbe domestication lies in the control of certain processes which occur at the molecular level. During the last thirty years there has been considerable progress in research in molecular biology; half of the Nobel Prizes in medicine awarded in the last fifteen years were for advances in that field.

The complexity of the technology and information required by those who work with microorganisms and enzymes might be compared to the complexity of knowledge that contributed so much to the authority of the atomic scientists in the 1940s and 1950s, which led to the control of nuclear energy—and to the manufacture of atomic weapons. It is this mass of accumulated knowledge in fields related to biology and chemistry—microbiology, molecular biology, genetics, biochemistry, organic chemistry, and chemical engineering—that allows us to predict an imminent revolution.

The ideal microbe is one that can produce an excess of a substance that has medical or industrial interest. Modern techniques issue from molecular biology, particularly from the work on cellular control that earned Nobel Prizes in medicine in 1965 for Lwoff, Monod, and Jacob, who discovered ways to stop or start at will the cellular machinery. The techniques of genetic engineering in this respect are also very promising. By transferring certain sequences of genes into bacteria that are easy to cultivate but are incapable of producing a given antibiotic or a given useful substance, one can change them into efficient producers of a given substance. Insulin, for example, can be made by the common bacteria of the intestine, *E. coli.* New antibiotics that enable us to combat more effectively bacteria that have become resistant to known antibiotics could also be made "to measure."*

Techniques of genetic engineering will also enable the production of biofertilizers by transferring genes that allow nitrogen fixation in symbiotic bacteria living in the roots of plants. Millions of people die of starvation because our industries do not know how to transform nitrogen (the 80 percent of the air we breathe that passes through our lungs unchanged) directly into ammonia or into nitrogen-containing molecules (the principal building blocks of proteins). In nature the nitrogen present in the air is transformed into ammonia by the symbiotic bacteria that live with vegetables such as peas and beans. The biological catalyst that effects the conversion of nitrogen to ammonia is an enzyme called nitrogenase. Its efficiency permits the annual conversion of 50 million tons of nitrogen (350 kilograms per hectare of vegetable crops, or about 770 pounds per half acre). In comparison, the fertilizer industry treats the same amount of nitrogen annually—50 million tons in 1973—but only by creating temperatures of 400° Centigrade and pressures of 200 atmospheres. It takes 20 million kilocalories to synthesize a ton of ammonia. Thus the

* In 1978 a team of scientists in California succeeded in a formidable task: an artificial gene coding for insulin was inserted in *E. coli,* which started to produce small quantities of the precious hormone. However, as promising as they are, these techniques represent new dangers for mankind. For this reason biologists have decided to apply the strictest security measures with respect to this type of genetic manipulation.

use of microorganisms in this type of process will not only have the advantage of supplying the human population with a nutritional supplement, it will also reduce the very high energy bill of the nitrogen fertilizer industry.

Microbes also know how to make proteins readily usable in human and animal nutrition. The decrease in the total quantity of proteins in the world—the result of poor harvests due to drought, insufficient catches of fish (particularly the anchovies used in fish meal), and the rise in the price of soybeans—is good reason for the growing bioindustry to produce proteins from microorganisms.

And the production of proteins by microbes offers other advantages: production is independent of agricultural and climatic conditions, the microbe biomass grows very quickly (this is particularly desirable in obtaining high yields), and production is not limited to the areas available for plant cultivation.*

Certain large corporations, such as British Petroleum (BP), used petroleum to grow yeast; others, such as ICI, use methanol. But one of the most interesting procedures is the one developed by the Bechtel Corporation and the University of Louisiana, in which microbes are fed abundant and inexpensive cellulose wastes: paper, wood pulp, sugar cane, animal manure, and corncobs.

In Europe 25 million tons of special food, made from six tons of soybean and fishmeal protein, is consumed annually by hogs. Proteins manufactured by microorganisms could represent two million tons of complementary food per year.

There is still much work for the microbes; we have hardly begun to explore the many avenues opened by the bioindustry. By taking advantage of the techniques of genetic engineering, fermentation, and automatic selection of bacterial strains, we shall be able to custom-produce microbes to perform special tasks such as the elimination of oil spills on the surface of the oceans, the production of biological light, and the manufacture of specialized pharmaceutical products.

The Domestication of Enzymes

The enzymes are the agents responsible for the specificity and the efficacy of the microbes. The domestication of these catalysts by the bioindustry opens the way to new forms of chemical transformation. We have already seen results in the pharmaceutical and food industries, in medicine, and in the use of new instruments for biomedical analysis.

---

* A 500-kg steer may yield ½-kg protein per day. The corresponding figure for 500 kgs of microbes is 5 *tons* per day.

In large part enzymes owe their catalytic properties to their "active site." The particular spot on the body of the enzyme where reactions occur at high speed depends on the tridimensional structure of the enzyme.* Thus the preservation of this structure is fundamental to enzyme activity. (This is why enzymes are so fragile.)

The ultimate goal of numerous researchers is to be able to synthesize artificial enzymes—or simply to copy, with the help of the appropriate molecules, the activity of the active site. The first synthesis of an enzyme was done in 1969 by research chemists at the Rockefeller Institute and by the drug manufacturer Merck Sharp & Dohme. Today there is automated machinery for making synthetic enzymes, but the mass production of custom-made enzymes for industrial or medical use has yet to be achieved. As nothing opposes this automatic synthesis, it should become an industrial reality in the near future.

The major conquest of the bioindustry and its most promising direction for development are in the field of *immobilized enzymes,* or insoluble enzymes. For some time the food and pharmaceutical industries have used "free" enzymes (enzymes in solution) in particularly delicate chemical reactions.

Using techniques borrowed from chemical engineering, we can now bind enzymes on plastic supports or enclose them in microcapsules. The activity of the enzymes is thus enhanced. Moreover, immobilized enzymes are reusable; they enable the reactions to take place continuously and over long periods of time.

A whole field of new applications of immobilized enzymes is opening up: the production of amino acids by treating a mixture of these acids with enzymes that selectively destroy one of two isomers, thereby isolating the desired one; the transformation of plant syrup dextrose into the fructose used in producing sweets and nonalcoholic drinks; and the selective hydrolysis of starch or cellulose molecules.

In the near future we foresee the manufacture of *biochemical electrodes* that will make possible biomedical measurements of high precision as well as the creation of instruments of analysis. Very small artificial kidneys can be made from immobilized urease. The treatment of tumors and metabolic disorders can also be studied with the help of immobilized enzymes.

The know-how gathered over the years in chemical engineering and in process control can be applied today to reactions implementing immobilized enzymes. The consequences of this transfer will be determining factors in the development of the bioindustry.

Instead of using immobilized enzymes, one can try to copy the activity of the catalytic site of the enzymes. This site is usually composed of a

---

* See the function of hemoglobin, p. 53–55.

metallic ion (iron, zinc, magnesium, or molybdenum) surrounded by a molecular chain containing specific groups of amino acids. The discipline which studies the catalytic activity of complexes formed by both metals and organic molecules is bioinorganic chemistry. We can expect great contributions to the development of the bioindustry from this new field.

Scientists have been able to accomplish the synthesis of bioinorganic complexes with iron, sulphur, molybdenum, and specific amino acids capable of simulating the nitrogenase enzyme that transforms nitrogen in the air into ammonia. Nitrogenase is so effective that just a few kilos—representing the total quantity of the enzymes present in all the nitrogen-fixing bacteria and algae in the ecosystem—is sufficient to transform millions of tons of nitrogen into ammonia annually. One can understand the interest in artificial catalysts; capable of functioning in very mild conditions, they offer radically new solutions to the energy and food crises.

Controlling Fermentation and Photosynthesis

Another very promising domain of the bioindustry involves the electronic control of fermentation reactions.

Fermentation is the oldest energy reaction among living things. From microbes to men, it provides either a part or the whole of the energy that serves to maintain the biological organization. The most primitive organisms survive and develop by fermenting (in the absence of oxygen) the organic substances they extract from their environment. The techniques of electronics, data processing, and automation now provide engineers with ways of helping microbes in their fermentation reactions and making their work more efficient, in the hope of accomplishing a number of tasks useful to society.

The basic method is to place the microorganisms in a fermenter and to furnish them with the elements they need to grow and develop, while automatically controlling the physical and chemical conditions of their environment. Computers help to determine and maintain the optimum conditions for the reactions: nutrients, acidity, concentration of carbon dioxide, elimination of wastes, and so on.

This system presents a curious and interesting symbiosis between man, computers, and microbes. Man obtains substances useful to his survival and development (such as drugs and proteins) and in return supplies the microbes with food and optimum conditions. The computers record, compare, and regulate the multitude of intermediate parameters typical of biological reactions, each of which possesses its own characteristics. This symbiosis is highly significant; it foreshadows the efficiency of the bioindustry of tomorrow and the perfection of electronic control necessary to the increase in productivity of our new microscopic slaves.

At a time of energy and environmental crises, one of the most useful biotransformations is the conversion of organic matter (from garbage, for example) to combustible gases, especially methane. The usual by-products of bacterial fermentation are carbon dioxide and methane (see p. 8), and the quantities produced are significant. In 1973 the one hundred largest cities in the United States produced 74 million tons of solid wastes. If these wastes had been converted into combustible gases, they would have provided a volume of methane equal to 3 percent of the total demand for natural gas in the United States. The biological production of methane can now be considered a complementary source of energy, one that offers to do away with significant quantities of garbage.

The domestication of photosynthesis reactions, toward which many laboratories around the world are working, will also have a determining effect on the development of the bioindustry. The ideal solution would see the production of energy-rich substances from the sun, carbon dioxide, water, and chlorophyll. Despite the rapid gains in knowledge in this area, however, we are still far from making a catalytic unit capable of reproducing the efficiency of the chloroplast in leaves. Intermediate solutions are nonetheless possible. Sugar cane, for example, produces the best photosynthetic yield of all known green plants. We should make greater use of it as an industrial raw material in the production of alcohol and ethylene and various carbon products. The hevea, too, is an efficient producer of carbon chains that could be used as a source of hydrocarbons.

Finally, chemical reactions which took place on the primitive earth billions of years ago can serve as models to the bioindustry. From simple gases (methane, ammonia, water vapor, hydrogen), under the effects of ultraviolet rays and in the presence of mineral catalysts, considerable masses of organic material were manufactured in the upper atmosphere and accumulated in the oceans. The first living organisms evolved from this reserve of material and food. Today the chemical industry, particularly in Japan, knows how to benefit from these "prebiotic" reactions in the manufacture of the raw materials used in making drugs. In this way the bioindustry could take advantage of the mild and natural reactions of prebiotic chemistry by channeling them toward the production of foods and pharmaceuticals.

Ecoengineering

The methods, reactions, and processes described above, like the bioindustry itself, are all part of a much larger body of techniques and skills. This domain will dominate the end of this century and the beginning of the next, just as mechanical engineering and then electronics have dominated the last fifty years. I call this assembly of techniques ecological engineering, or *ecoengineering.*

Ecoengineering is much more than ecological development, management, or planning. Reaching beyond the management of nature, it recognizes the symbiotic nature of the relationship between human society and the ecosystem, wherein each uses the other to their mutual benefit.

Ecoengineering, with the help of new methods like energy analysis, will enable men for the first time to control consciously the energy circuits of the ecosystem for the good of man and nature.

Like doctors or surgeons working in the very interior of the organism, we shall be able to reestablish the great feedback loops of reward and reinforcement on which the "economy" of nature is built. We shall have to close in, reconnect, and even "naturalize" the chains and networks of the socioeconomic system (such as those that eliminate wastes and produce food). We shall be able to develop new bacterial strains capable of helping us to effect a more efficient recycling of used materials and elimination of waste. We shall have to achieve the large-scale transformation of nitrogen into ammonia in order to feed the world population. We shall have to modify climates locally to provide new areas for cultivation and to help nature readapt to the aggressions to which we have subjected her.

With the advent of ecoengineering, the dangerous experiments of those of us who are sorcerers' apprentices will cease. Only then shall we be able to develop a partnership between man and nature, the basis for a new postindustrial economy and society that we shall have to create from scratch.

Four

Information and the Interactive Society

Information, too, is energy, a particular kind of energy that releases and controls power. The close relationship between energy and information came to light when it was understood that energy had to be spent in order to acquire information and information had to be used in order to collect energy and put it to use. Every bit of information has to be paid for in energy, and every increase in energy must be paid for in information.

Information would have remained a qualitative concept of little interest if it had not become possible to measure precisely the *amount of information* contained in a message passing through a transmission line. This ability to measure information, achieved in the late 1940s, led to a veritable revolution in mathematics, physics, and electronics. Its impact has been particularly marked in cybernetics, data processing, and telecommunications.

One of the most profitable ways to understand the concept of information and the consequences of the revolution it fostered is to take (as we did for energy) a position that enables us to observe "through the macroscope" the role that information and communications play in society. This leads first to a review of several important points concerning communications, the measurement of information, and the relationship between information and entropy. Then, following a brief history of communications, we shall come to a discussion of the conditions and possible consequences of the appearance of an interactive and participative society founded on telecommunications and what I call "society in real time."

1. SUPPORTS OF COMMUNICATION

There is a profound difference between matter and form. Matter seems immutable; it conserves its shape and does not change. It is form that changes and modifies itself. This difference in nature was illustrated by Aristotle in his famous example of the brass statue.

Aristotle introduced another distinction—of perhaps greater importance—between the two meanings of the word *information.* On the one hand, information is understood as "the acquisition of knowledge" (one *becomes informed* by the act of observing an object or observing nature). On the other hand, information denotes "the ability to organize" or "creative action" (one *informs* matter by the act of giving form to an object—as the sculptor does with clay).

For the moment, we can define information simply as *the content of a message capable of triggering action.* Later we shall consider the more precise definition proposed by the theory of information.

Communication is the exchange and circulation of information in a network that connects transmitters and receivers. Information is sent from a transmitter to a receiver in the form of a *message.* A message is composed of signals, signs, or symbols assembled according to a *code.* There are coded messages, communications in Morse code, and hereditary information enclosed in the DNA molecule in the form of the genetic code. An elaborate collection of codes and messages forms a *language.* A message is coded at its source, then sent by means of a carrier (Fig. 74).

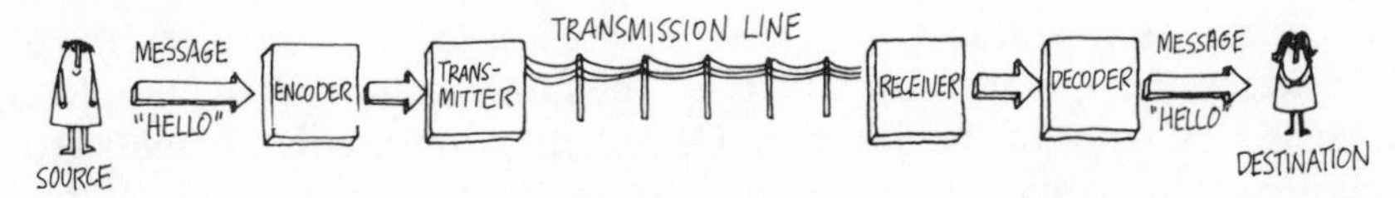

Fig. 74

Whatever its nature (radio waves, wires, laser beams), the carrier is called a *transmission line* or wave. At the other end of this line the message is decoded and transcribed into information that has meaning for the person to whom it is addressed. But in order for the recipient to recognize and use the information, there must already have been information *memorized* that can be compared with the message just received. A final and important point is that disturbances occurring in the transmission line (the "noise") can alter the message and change its meaning.

Measuring Information

The significance of information varies from individual to individual. The information that it is going to rain will have entirely different effects on a vacationer hoping for sunny days and on a farmer threatened by drought. In its most current sense, information is a new fact, an insight, or knowledge newly gained from observation. Information can be stored in one's memory or in libraries, where it serves to support effective action.

Thus it would seem to be impossible to measure information. In order to do so, all reference to its subject matter would have to be ignored and only the specific form of energy passing through the carrier considered. This particular "information" has a much more restricted sense than current usage assigns to it. But its definition has made it possible to arrive at a quantitative expression that is indispensable to improved communications and the future of data processing.

The measurement of information is the result of a remarkable convergence of independent efforts undertaken in the late 1940s by telecommunications and servomechanisms engineers, by mathematicians, by statistical mechanics theorists, and by physicists. The theory of information had its beginnings in this work and culminated in Shannon and Weaver's *The Mathematical Theory of Communication.*

The various researchers had been led, as a result of their experiments and findings, to make a number of concluding observations.

Information that travels along a transmission line degrades in an irreversible manner. In this respect information is analogous to energy, which degrades into entropy. If, for example, one takes the mold of a statue and casts another statue from it, then makes a mold of the second statue and casts a new statue from it, it is quite possible that after twenty such successive operations the form of the final statue will be completely different from that of the first one. Or one may make enlargements from a photographic negative and from them make new prints. The slightest scratch will irreversibly affect the original information.

Energy must be used in order to transmit information. The energy support of information is represented by light radiation, sound waves, the electric current in a telephone line, and a bee that carries pollen from flower to flower. As this energy weakens and becomes dispersed, it must be channeled and amplified. Finally, the greater the precision of measurement, the larger will be the amount of energy expended.

In order to avoid the degradation of information and improve the quality of transmission, one must first measure the quantity of information contained in a message.

To define properly what a certain quantity of information represents, one must put oneself in the situation of an observer trying to obtain information about a system that he does not understand. This system could be made up of the number of possible answers to a question, the number of solutions to a problem, or simply a pack of cards spread out on a table.

Obtaining information about an unknown system can lead the observer to reduce the number of possible answers. Total information could even lead immediately to the only possible answer: the correct one. Information is therefore a function of the relationship between the number of possible responses *before* the reception of the message (P_0) and the number of responses possible *afterward* (P_1).

Consider a simple example. The unknown system is a pack of thirty-two cards; what chance is there of drawing a card named in advance?

This question introduces an uncertainty, and this uncertainty is measured by a ratio: the number of favorable choices to the number of possible choices. This is called the probability of drawing the correct card. As there is only one favorable choice (the card named), the probability here is one chance in thirty-two.

Now how does one measure the amount of information acquired by drawing a card? *Before* a card is drawn there are thirty-two possible choices, all with the same probability (P_0). After a card has been drawn, two situations are present:

If you have drawn the right card, only one answer is possible and you are holding it in your hand. The amount of information obtained is a function of the ratio 32:1, and the information is total.

If you have drawn the wrong card, thirty-one possible responses remain. The amount of information is a function of the ratio 32:31, and the information is partial.

The information obtained in the first situation fully resolves the problem by reducing the number of possible chances to one. In the second situation it diminishes slightly the number of possible chances. Here it reduces the denominator of the fraction P_0/P_1; thus the ratio increases, and so does the information. This leads us to conclude that information increases when uncertainty diminishes—because uncertainty indicates the lack of information that one has of an unknown system.

Finally, in order to measure information and define unities, one can adopt two conventions. One can choose to define information in a subtractive way rather than by a ratio, since information is the *difference between two uncertainties* (before the message and after). For the ratio P_0/P_1 one substitutes a subtraction of the logarithm.* The second convention involves the most convenient and the most used code for sending a message, the two signs 0 and 1, which can also stand for yes and no. This leads to the adoption of the binary language and logarithms of base 2. Applying these conventions, the amount of information in a message is measured in "bits" (an abbreviation of *binary digits).*

Now we can answer the question, what is the amount of information acquired by drawing a card? The amount is 5 bits (in base 2, the logarithm of 32 is 5, or $32 = 2^5$).

---

* The logarithm of a number is the power to which the base must be raised to obtain the number. In base 10 the logarithm of 1,000 is 3, because $1{,}000 = 10^3$. In base 2 the logarithm of 16 is 4, because $16 = 2^4$.

Thus information seems like an abstract entity, objective, devoid of human meaning. It is easier to represent a given amount of information by comparing it to material units circulating in a conduit, like molecules of water in a pipe. The capacity of the pipe is limited by its size, and the same is true of a transmission line. Some lines—a standard telephone line, for example—cannot carry more than 1,200 bits per second. This amount of information is entirely independent of the significance of the message, which could be a song, racing results, or stock market quotations.

Information and Entropy

Any information that results from an observation, a measurement, or an experiment, and that tells us only what we already know, produces no change in the number of possible responses; it does not diminish our uncertainty. The lower the probability that a message or an event will occur, the greater is the information carried by that message. The information obtained by drawing the correct response the first time ($I = 32/1$) is the inverse of the *probability* of obtaining this response before the drawing is made, or before the message is received ($P = 1/32$). Probability and entropy are related by statistical theory (see p. 102). By bringing together the different mathematical expressions, we can see that information is the *inverse* of the entropy of the physicists; it is the equivalent of an antientropy. The term *neguentropy,* negative entropy, has been proposed to identify this important property. Information and neguentropy are therefore the equivalents of potential energy.

The alliance goes further. By choosing suitable constants and values, one can express information in thermodynamic units and relate it directly to entropy. We can then calculate the smallest expense of energy needed to generate one bit of information. To obtain an amount of information equal to one bit, we must degrade in entropy a very low but finite and therefore significant quantity of the energy of the universe.

This important finding has led physicists like Léon Brillouin to generalize Carnot's principle in such a way as to express the indissoluble relationship that exists between information acquired by the brain and the variation of entropy in the universe: Every acquisition of knowledge based on an observation or a physical measurement obtained with the help of an instrument uses energy in the laboratory—and therefore some of the energy of the universe.*

Consider an example. The reading of this page involves several elements: the text (printed in black on the paper), a source of light (natural

* Inversely, the brain creates information and thus can decrease entropy. We will see this in the following chapter, on time.

or artificial), the eye, and the brain. The lamp is the source of neguentropy. It emits a flow of light that is refracted on the succession of black and white segments of the printed words and modulates the light beam that strikes the eye. The eye receives the message and the brain decodes and interprets it. Thus the reader's brain has acquired information. But this must be paid for in energy: the watts of the lamp in exchange for the 24,000 bits of information on the printed page.

The History of Communications

The history of communications begins at the molecular level. A large part of the information on which communications between molecules depends is built into their shape, their principal support of signals, codes, and messages. Molecules are "information individuals" who carry, inscribed in their morphology, what they are, what they do, what they know, and what "memory" they have that enables them to "recognize" other forms.

The cell maintains its organization, its complexity, and its coordination by means of a complicated network of intermolecular communications. The enzymes, situated at the nodes of these networks, screen the molecules and control the flows of information, thereby permitting the rapidity and efficiency of the fundamental reactions of life.

The DNA molecule, the support of genetic information, illustrates perhaps better than any other biomolecule the basic principles of communication. (In recounting its role, we italicize those terms common to biologists and communications engineers.) Genetic information is stored in the form of a molecular *code*. It is *transcribed* in RNA molecules, the *carrier* that transports *copies* of this information from the nucleus to the cell. By the action of ribosomes and molecules of *transfer* RNA, the information is *translated* into protein molecules. With an "alphabet" of twenty amino acids, the cell manufactures thousands of different proteins in the same way that we compose thousands of different sentences with the twenty-six letters of our alphabet. During the replication of DNA molecules, disturbances from the environment (the *noise* on the transmission line) introduce "errors" that change the meaning of the *messages;* these modifications are *mutations*.

Chemical communication by means of the shape of molecules is the oldest system of communication used by living systems. The molecule signals are not only responsible for control and regulation of internal activities of the cell; they cross the membrane, circulate in the neighboring milieu, and send signals to other cells. The behavior of bacteria, yeast, algae, and protozoa depends on the chemical messages they exchange among themselves or with the environment. Microorganisms know how to recognize and avoid poisons and how to guide themselves toward nutrients. When a blood cell under the microscope is suddenly killed by a laser beam, one sees the freed chemical substances immediately attract white cells, which rush toward the dead cell in order to absorb it.

Certain cells living in cultures *in vitro* synchronize their activities, their movements, and their pulsations through the emission of chemical substances that act as coordinating signals (this is seen in a spectacular way in cultures of heart cells). And more advanced means of communication already occur in unicellular organisms. Numerous microorganisms capable of photosynthesis have a "visual spot" made up of light-sensitive molecules—virtually a primitive eye—that enables them to move toward a source of light. Little muscular fibers move vibrating ciliae, which are generators of movement and thus a means of communication. In most cells there are microtubules that resemble the pipes and channels found everywhere in nature and through which liquids and materials pass.

The integration and differentiation of the cells of the tissues and organs at the heart of the organism lead to the diversification of means of communication. The support of genetic information and the coordination of communication and cellular regulation rests with DNA and its performing agents, the enzymes. But when a short response time is required, internal and external communication are achieved through the nervous system and the hormones (see p. 41), which permit rapid reactions to stimuli from the environment.

Chemical communication has its ways, too. Odors given off by insects and animals, toxic products, poisons, venoms, plant alkaloids, the scents of plants—especially flowers—guarantee the regulation of natural equilibriums and the maintenance of the entire organization. The power of pheromone, a chemical substance used by insects for communication, is so strong that a single molecule picked up more than half a mile away by the antennae of the silkworm butterfly will lead it to the female.

With vision and hearing comes an explosive diversity of communication. In a world bright and variegated with flowers, fish, the plumage of birds, and the coats of animals, one hears the responsive songs, cries, and calls. A multitude of phosphorescent spots illuminate the depths of the oceans and the darkness of the night. Each sign has a precise meaning in a given environment; this is a form of social communication.

Among the more highly evolved animals, auditory, visual, and olfactory communication can be expanded through posture, the position of the limbs, and in primates facial expression. The marking out of a territory by odors—a very old form of communication—or by touch (the rubbing together of insect antennae, delousing among monkeys) emphasizes the effects of other forms of communication and helps to increase the variety and stability of the different ecological niches.

Communication between human beings must be considered separately because of the importance of language. This does not mean that man is excluded from using nonverbal—visual, olfactory, tactile—forms of communication; essentially animal forms of expression, they can set in motion an infinite variety of behavior patterns. But technical and social progress is founded mainly on the creative power of language and the logical thought that derives from it.

The major phases of communications development have followed an accelerated pace from the drawings of prehistoric man to papyrus manuscripts and to electronic impulses and television. For the speed of evolution has depended increasingly on fluid, adaptable, "nonmaterial" systems such as printing and now electronics. Following the appearance of language and the generalization of the oral tradition, the advent of writing allowed information to be expanded and stored at small energy cost. The practice of copying manuscripts, the invention of printing, and the creation of libraries exteriorized one of the principal functions of the human brain—memory—by freeing the prodigious power of the amplification of information. One characteristic of every social organization is to achieve—in as short a time as possible in proportion to the organization's complexity—the multiplication and spread of the total mass of existing information, with as little energy cost as possible.

The true telecommunications explosion began when man learned to code and transmit information by wire or high-frequency waves. With the telegraph and the telephone, radio and television, sound and image conquered oceans and mountains, encircled the globe, and reunited men in the "global village" so dear to Marshall McLuhan. The letter, the telephone, and the shortwave radio allowed only bilateral communication or, at best, communication among small groups. Radio, television, newspapers, and magazines reach a large number of individuals, but those people are deprived of the feedback control of information.

In the "global village" communication no longer depends solely on written, spoken, or audiovisual information. There is a world of signs and symbols of infinite complexity, and the strength of their messages is as real as the printed word or the televised picture. Dress, social behavior, the signs of the purchase and ownership of material goods such as a car or a home, and art, music, and sports, too, are means of communication that can assure the integration and the complementary differentiation of the individuals within a social organization.

Today, in the linking of computers and telecommunications networks, we are witnessing the assembly of a veritable public utility for information. Such a network will represent the most elaborate stage of the integration of the various systems of communication from the molecule signals of the bacteria to the nervous systems of man and society.

The foregoing "natural history" of the role of information and communications in biological systems and animal and human societies necessarily leads us to the question of the next step in the evolution of communications. Will the planetary system under construction be the "nervous system" of our societies? Will it be the material support of the noosphere, the sphere of the mind that Teilhard de Chardin saw as the successor to the biosphere, the sphere of life?

One process appears to be irreversibly active in most developed countries: the increasingly closer integration of the human brain, telecommunications systems, and the computer. This process, if it continues, may well be the support of a new form of social organization. Will it be an interactive and participative society that respects individual initiative and the pluralism of ideas? Or will it be a caricature of society approaching that described by Orwell in *1984*?

The speed of evolution and the impact of telecommunications systems are such that it seems to me useful to discuss now the conditions and the consequences of the possible future of a new form of social organization, "society in real time."

The expression "real time" comes from the vocabulary of computer programmers. We say that a dialogue or an interaction (between man and computer, for example) develops in real time when the information coming from the environment is treated as it arrives. This idea can be generalized: every action that involves decisions and deadlines happens in real time when the information that is the basis for the decisions reaches the decision centers *before* the deadlines. The standard "real time" is the maximum time allowed so that information involving a decision can reach a receiver before the decision is made.

This maximum time varies considerably: several microseconds in the case of a computer controlling the release of a rocket; several seconds or several minutes in the control of assembly lines in an automated factory; several months in the case of social systems. In daily life the concept of real time is linked to concepts of interaction with other persons or with machines. Interaction makes possible the immediate reception of information or signals (movements, facial expressions, intonations of the voice) by which behavior and decisions are modified. The concept of real time is also linked to that of "live" events presented by radio and television broadcasts, which allow participation in far-off events.

Descending and Ascending Information

The birth of society in real time will result from the evolution of two complementary forms of communications systems. One of these evolutions is at a more advanced stage than the other, and this creates an imbalance whose sometimes dramatic results are now being felt.

The two evolutions involve the continuation in society of two fundamental actions of the individual conscience: *observation* (acquiring knowledge, informing oneself) and *creative action* (organizing the world, informing matter). In the first case, all acquisition of knowledge is counterbalanced by an increase in entropy in the universe. In the second case, all creation of new information by the human brain contributes to a decrease in entropy locally. Daily experience shows that the first mode of activity is considered easy, requiring little effort; the second is considered more difficult, more demanding.

In a similar way, society has endowed itself with a system of communications based on the rapid dissemination of information. From the top of the pyramid that is the form of every social organization down to its base, there is a system of *descending* information.

This system represents the large-scale transposition of the act of observation or the acquisition of information by the brain. It is given form by the well-known mass media (books, newspapers, radio, cinema, and television), which carry descending information to all parts of the earth. Its evolution has been rapid and its activity explosive, for the copying and distribution of information can be done on a grand scale at a minimum energy cost.

The other system of communication has only gradually been put to work. Still far from achieving the effectiveness of the first system, the second system is principally one of sending information back to decision or broadcasting centers. This is *ascending* information: individual actions and personal participation or contributions to the functioning of an organization or the greater social system. It is the transposition to the collective plan of creative action that each person performs at his own level.

This system (we call it ascending for symmetry) is made up of all the everyday forms of representation and participation in the life of society: the vote, elected representatives, political parties, production committees, labor unions, consumer institutes, public opinion polls. This is the "response" of individual members of society to politics, to government programs, to the management of a company, to the mass of goods and services provided by industry.

The slow pace of its operation can be explained by the high price that one must pay in information (the education required at every level) so that each individual can participate effectively in the organization and development of society. Every creation of a new organization (the equivalent of potential energy known as neguentropy) must be counterbalanced by a significant expense of information (Fig. 75).

In addition to the systems of descending and ascending information, there is the entire network of horizontal communications, from person to person or from person to machine—first by means of mail and the telephone, then with the interactive electronic systems that are just coming into their own. The integration of these three systems of communication provides a rough sketch of the infrastructure of society in real time.*

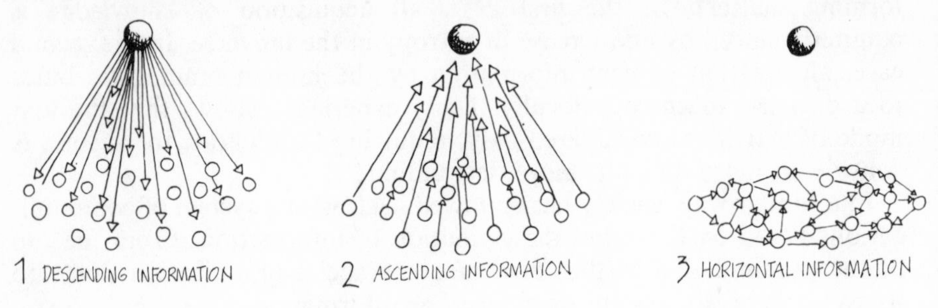

Fig. 75

2. THE NEW INTERACTIVE NETWORKS

The technology of communications is well known and readily available. It has arrived at a stage of development, especially in the United States, where it is possible to talk about the coming of a public utility for information, the embryo of society in real time. As a complement to other large-scale public utilities for energy and transportation, the new information utility will probably have a more significant impact on the organization of society.

Yet the real problems are not technical; they are political and economic problems. We have no idea of the wide-ranging consequences of an expansion of the contacts and interactions in real time among the inhabitants of a country, in their homes or at their places of work. We have no idea of the effects of their *selective* access to information—to cultural activities or to entertainment. Who has been able to measure the social and economic impact of the telephone? How could we foresee the impact of computerized information and communications systems on transportation and travel, on the organization of large cities, on the working habits of the population, and on education? Will an "interactive" society lead to the sense of participation of its citizens? Without receiving prompt feedback within a reasonable period of time as a response to his actions, an individual loses all feeling of participation in the operation of the system to which he belongs; he tends to become passive and disinterested in the organization on which he depends. One of the forms of "social

* The circuits of ascending and descending information exist only where there is centralized power. In a decentralized, interactive, and participative society, the wealth of interaction in real time is founded on the diversity of exchanges between individuals. Decentralization of power comes about naturally through a growth in individual responsibility and in pluralism.

malaise" can be described, as Jacques Attali has said, as the feeling of being *left out of power* that is felt by every citizen deprived of a real means of participating.

We must no longer let ourselves be carried away by the perspectives that communications technology has opened up. A computer terminal in the home or interactive networks of cable television will be expensive, and it is not certain that they are necessary or even wanted. Will the social cost be justified? How can we distinguish, in the maze of electronic gadgets constructed by telecommunications engineers, those that will have real advantages for society and the individual?

One of the major differences between the time in which we live and that of the great technical breakthroughs of the first half of the twentieth century is that we do not have to endure the effects of poorly planned and controlled technology. We are now able, perhaps for the first time, to prepare—with full hearings—for the introduction of new technologies into our lives in a manner that is in the best interests of man and society.

Another difference is that instead of serving prestigious operations undertaken for political ends, communications technology can render immediate service to all citizens in a form that they will be able to understand and appreciate. But once they have decided to accept the installation of such networks in principle, will they be ready to assume the costs?

When the means of communication are installed, their services will expand. The more the services expand, the less the operating cost of the network will increase. Yet no service will expand at too high a cost once the network is installed. Again we are trapped in a vicious circle: which should come first, the telecommunications network or the service that makes use of it? We might say both at once, first in an embryonic way, then in a more complex way as a result of continuing evolution. Political desire, public pressure, or the urgency of a situation can accelerate the process. For the moment none of these services can be justified on economic grounds alone, yet we sense that their coming is inevitable.

There is as much promise as there are dangers in the advent of public information utilities. There is the promise of a more humane society, one that is less centralized, that will bring people together and profit from their interactions. At the same time there are the dangers of mass manipulation, of the infringement of individual privacy, of a new form of social inequality based on preferential access to information. To measure the magnitude of the revolution that is brewing and its potential impact on our daily lives, we must consider now the technical support of the new communications systems and above all the services that rely on the networks of descending and interpersonal information. Then we shall consider the problems posed by ascending information, the support of participation in real time.

Communications Hardware

Knowledge is power, the proverb reminds us. Heretofore the control of information—and hence a share of power—has been in the hands of small political groups and private businesses. Now modern communications technology in theory offers the possibility of a complete redistribution of power. For the first time, information carried by the transmission lines can be controlled *by the receiver* rather than by the source.

To understand this unprecedented change and its social repercussions, one must compare present-day communications technology with that which would replace it in society in real time. The principal forms of the mass media today fall into two large groups: the storage media for texts, images, and sounds (books, newspapers, films, recordings) and the transmission media (radio, television, telephone).

Closely related to the storage media are the powerful duplication media, which allow the production of large numbers of copies of books and newspapers, and the distribution networks (bookstores, newsstands, cinemas, record shops). Information can be delivered to homes by mail subscription as well.

Radio and television, the transmission media, also act as duplication media by sending the same information simultaneously to a large number of people. They make it possible to transmit and broadcast audiovisual storage media—records on radio, films on television. However, the selection of programs and the hours of broadcasting remain under the control of the source.

The only primary large-scale transmission medium controlled by users is the telephone, but it is not generally linked to the mass storage media. Like the other interactive media—letters and shortwave radio, but not the new citizens' band network—it allows only bilateral communication. One of the few opportunities the user has to exercise direct control over transmission from storage media is to go to a bookstore or a newsstand and select a book or magazine.

The situation is entirely different with the new electronic storage and transmission media. These systems have an electronic or computerized data bank, a transmission network, and computer terminals in the users' homes for selective access to information. The data banks are either magnetic disks containing up to 800 billion bits of information (the equivalent of 100,000 books of 400 pages) or microfilms stored in an access system that can be controlled by computer.

The transmission networks use telephone lines or cable television, and their impact is directly related to their transmission capacity. What are their limits? Consider some examples of transmission range.

One page of this book contains about 3,000 characters or 24,000 bits of information. A fast reader can read this page in a minute at a rate of 500 words per minute or about 400 bits per second. In comparison, the capacity of a telegraphic transmission line is about 75 bits per second. A telephone line carries an average of 1,200 bits and as much as 9,600 bits per second. If the information is transmitted in digital form, 60,000 bits per second can be sent.

The newest communication systems are the microwave transmission networks and the coaxial cable. Using relay antennas, a microwave transmission network can carry as many as 100 telephone communications simultaneously at a speed of 70 million bits per second. Services using these networks are already connected with banks, hotels, airline reservations agents, and computer services. The coaxial cable, built around a conductor in a hollow tube, can carry 10,000 telephone communications and 700 billion bits per second. As the basis of cable television, this network creates "wired cities," connecting users and relay stations and ensuring two-way information between subscribers and central stations and among the subscribers themselves (Fig. 76).

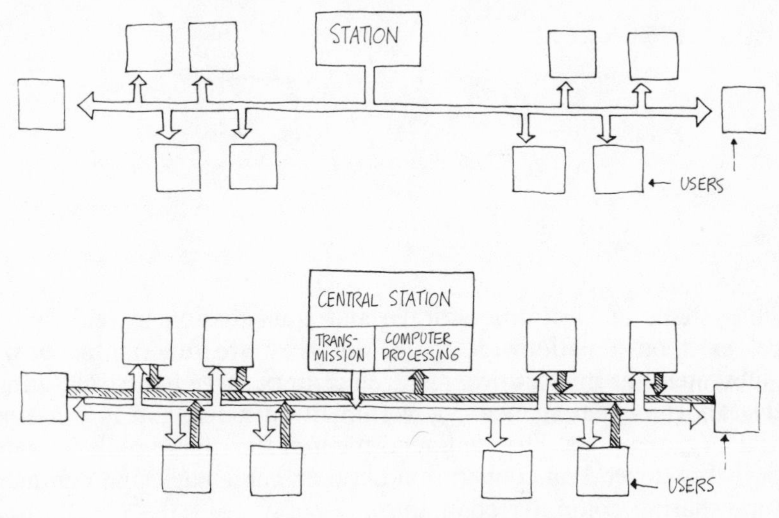

Fig. 76

Communications technology has still other systems in reserve—satellites, wave guides, and optic wave guides using laser beams. The wave guide is a hollow tube in which 250,000 telephone conversations can travel simultaneously—a flow of information of 15 million bits per second. The theoretical capacity of the laser reaches tens of millions of simultaneous communications; it will probably surpass all present and future needs of society.

Today the most common terminals found in homes are the telephone and the television set. Their newest versions include the *touch telephone,* which allows communication with computers, and the *videotelephone* and *interactive cable television,* which are competitive systems.

In the future the communications terminal in the home will probably look like a combination of television set, telephone, and teletype. It will function at the same time as a library, a news magazine, a mail-order catalogue, a postal service, a classroom, a theater, and a telephone inquiry service. Time-sharing computers connected to transmission networks will ensure the selection of information, the control of communications between subscribers, and the storage of information in data banks (Fig. 77).

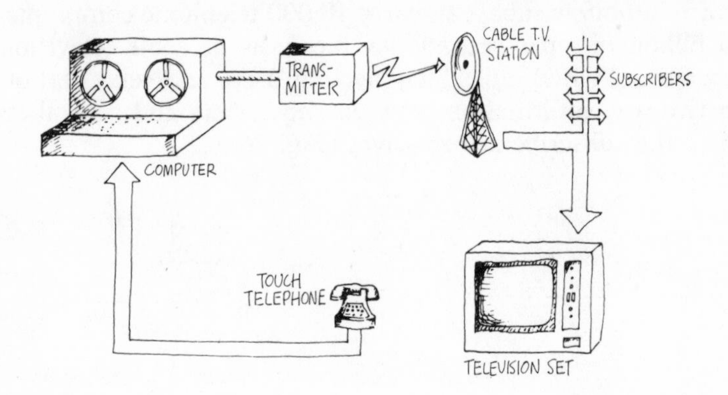

Fig. 77

Such systems of electronic storage and transmission in real time do not yet exist on a nationwide scale. But they are functioning now in such subsystems as universities, research centers, large industries, administrative and financial agencies, government agencies, and international scientific organizations. The commencement of their large-scale operation will probably depend on cooperation between cable television companies and time-sharing computer companies.

Services in Real Time

The wired city is becoming a reality. There will be 30 million homes in the United States in 1980 that will have cable television. In France new cities like Créteil and Cergy-Pontoise will be "wired." In fact all cities are already wired with electricity and telephone lines. The coaxial television cable, because of its two-way capability for transmitting information, opens the way to a new era of services. And there is more than cable television; the interaction among network subscribers and between subscribers and central stations can be as readily accomplished by expanding the telephone network and using videotelephones.

Without distinguishing between the various specific uses of cable television and videotelephone systems, one can still offer a glimpse of the kinds of services that society in real time will be able to offer. In describing them now as though they already existed, I am also raising the question, is this what we want for tomorrow?

Selective access to information. Any subscriber to cable television has the same advantage: a turn of the knob permits the choice of forty channels. Another service lets one communicate by touch telephone with a computer at a central station and call for the news or programs that one wants from a television station.

Subscribers at home can also request information from data banks, whose contents are then presented visually in graphs, photos, and films. Subscribers have instant access to legal, administrative, financial, and technical information and sports data. They can go through archives and study rare documents; they can visit museums and exhibitions.

Doctors and engineers who subscribe to selective information dissemination services are alerted to the appearance of publications in their area of interest. A custom-made newspaper that corresponds to the profile of interest of the subscriber is delivered to his home. Thus information becomes more useful, more varied, more individualized.

Visual communication. Communication by videotelephone clearly goes beyond the simple amusement of seeing one's correspondent on the screen. Conferences can be held between several people separated by distances of hundreds of miles. Relatives can "visit" patients in hospitals or prisoners in jails. Students can attend lectures or special courses. Doctors can examine their patients as they consult their medical files (x-rays, electrocardiograms, encephalograms, histologies). Lawyers and legal, financial, and technical advisors can work with their clients over the same files. Private conversations can be held with marriage and family planning counselors. Businesses can conduct preliminary interviews with job candidates. News editors and reporters, copyeditors and authors can work together on layout and editing.

When linked with computers, the visual communications network will offer an expanded range of services, including computer-assisted instruction in which tests, problems, and exercises are the framework of a more individualized instruction. The management in real time of bank accounts has become necessary because there are now systems of payment that do not involve checks or cash. The bank's computer answers questions by displaying its response on a television screen or "talking" on the telephone.

Interpersonal communication. Information on the most diverse fields—for which there is supply and demand—is stored in data banks and kept up to date as subscribers communicate new information. As if in a kind of electronic classified advertising service, the computer compares the characteristics of each offer and each request and puts the appropriate parties in touch. This is the computer matching system; it brings together ideas, situations, and interests.

In place of impersonal and inefficient mass contact, selective matching modifies the quality of personal relations by increasing the probability of cross-fertilization of ideas, the comparison of original efforts, and the mobility of people and ideas. Computer matching has direct application in school, university, and vocational guidance, in employment agencies that embrace several cities, and in transfers of technology between nations.

Putting people in touch through selective matching can lead to a better rate of utilization of common goods. People working in the same area of the city and living near one another can use the same car to get to and from work. Computerized car pools have had varying degrees of success in the United States and in Europe, depending on the area. Yet the advantages of this simple idea are obvious; it could well be extended to other fields.

In the more distant future, network subscribers will be able to communicate through decentralized data banks accessible to all. These banks will store publications, employment and product offers, and ideas. Every individual will be able to explore the bank selectively; when he finds what he wants, he will be put in contact with those whose ideas, tastes, or activities match his own.

Control of the city's functions. With interactive networks, the city becomes more and more like a living organism. The wired city assures its residents of protection against fire and theft, for fire detectors and burglar alarms are linked directly to surveillance and emergency services by coaxial cable. Gas and electricity services will take their meter readings directly from the home. The police will be able to hold lineups of suspects, compare fingerprints, and study stolen cars and other objects at distant locations. Detectors and television cameras placed along roads and at intersections will report to computers the necessary information for controlling city and highway traffic in real time. The automation of traffic lights and alternate routes in the event of tieups or bad weather conditions will ensure an improved flow of traffic. Automatic identification of moving vehicles in certain zones will make it possible to provide traffic information and to control traffic lights in order to allow emergency vehicles the right of way.

Even the miniature communication systems dear to science fiction writers have become a reality. Private cars already have radiotelephones; in some countries there are methods of locating and paging people by beeps—a system first used to contact doctors in emergency cases. These systems function not only within large organizations but as Nationwide Paging Systems. Instantaneous person-to-person communication by means of a wristwatch transmitter-receiver is technically feasible, even at great distances.

The dialogue between computers. Computer networks are being connected with one another. The ARPA network joins the computers of thirty American universities with the help of small computers capable of translation in real time. Veritable "ganglions" at the nodes of a nervous system, they make possible the translation of messages from one computer language to another. This network now extends to European computers, giving them virtually instantaneous access to all the libraries of specialized programs now in operation on the campuses of thirty American universities.

Teleconferences, already widespread since they are offered as a supplementary service by telephone companies, will now benefit by the interconnection with computers. The Institute for the Future in the United States has developed a system (FORUM) that allows experts to communicate in real time or in delayed time with other experts on a given subject. All benefit from the computers' information-processing capabilities and from the specialized information stored in their data banks.

Social Impact of Services in Real Time

Over the past ten years one of the favorite topics of the futurists has been the potential of new communication systems to substitute for human travel. To move information through wires instead of moving people over highways appears to be more efficient: it saves energy as well as time.

Few studies have been made of the relationship between transportation and communication, but the energy crises have given the matter high priority. A reduction in the amount of professional travel and the daily commuting of city dwellers would not only cut down on fuel consumption, it would reduce pollution, noise, and tension in large cities. Even if this substitution affected only 18 percent of the travel in a city (as shown in studies made in 1972), it would have considerable influence on living conditions in the large metropolises.

This does not mean that we are headed for a society in which people travel only for pleasure or leisure. Some kinds of work will always require travel. It is difficult to imagine a chef preparing meals or a barber giving a haircut by closed-circuit television. Similarly, one cannot participate in sports, acquire a suntan, or breathe the forest air by using a videotelephone (fortunately!). On the other hand, practically everything done in an office—reading, writing, dictating notes or correspondence, telephoning, attending meetings—could be done from one's home. True, nothing can replace personal contact, and nothing will prevent occasional traveling to call on a customer, to sign a contract, to visit a factory, or to evaluate people.

The bulk of the communications that constitute business life will be carried on more and more by means of videotelephone networks, cable television, and teleconferences. Calculations made at Cornell University in 1973 and cited by Edward N. Dickson in a report on the impact of the videotelephone attempt to evaluate its costs compared to those of travel and electronic communication.

In a time of energy crises these comparisons are very interesting. Eight hours of transatlantic travel in a Boeing 747 for the purpose of meeting someone personally uses eight times more energy than a videotelephone conversation of the same duration. For short distances, the energy in five liters of gasoline can fuel a car for about 30 miles or provide 66 hours of uninterrupted videotelephone conversation. At present the substitution of the videotelephone for travel holds no interest on economic grounds, but telecommunications specialists agree that in the long run travel will become less efficient and more costly. The continuing replacement of some kinds of travel by electronic communications will probably have a pronounced effect on the organization of large cities. As a result of decentralization, the metropolises will split up into villagelike communities whose inhabitants will work at home. Such an evolution will lead to a "new rural society."

Society in real time will witness a revolution in education. Through selective matching and person-to-person communication, people of all ages and all social levels will be able to benefit from individualized education.

Interactive networks will bring about further developments in service activities. Industrial civilization is founded on the principle of mass production; the beginnings of informational civilization, however, rest on selective production and destandardization. The success of products made by craftsmen and the number of magazines produced for small special interest groups are harbingers. Bringing people together by means of visual communication will lead to the creation of a multitude of new services and accelerate the "dematerializing" of the economy that is already under way.*

* This willfully futuristic description of services in real time has tried to show toward what the explosive development of telecommunications and data processing can lead. It is not a technological forecast. We must also take notice of the other extreme view, shared by sociologists and architects

3. SOCIAL FEEDBACK

One of the most important advantages of the new electronic information systems is the possibility of the feedback of information to decision centers. Without feedback loops there can be no efficient participation, no interactive society.

The feedback of information at all levels of the social organization (businesses, cities, states, governments) represents a great loop of cybernetic control that I call *social feedback.*

Without control loops a social system under "direct command" is nothing more than a dictatorship; only with the installation of control loops can the system evolve toward democracy. Today the effectiveness of regulatory systems and traditional participation—and above all, the length of their response time—does not satisfy the demands of a rapidly growing society.

The oldest forms of social feedback are probably the applause and the catcalls of a crowd, but the most widely used form is clearly the vote. We all know the limitations and weaknesses of the vote: discontinuous participation, delays in tabulation, excessive simplification of choices, the inability to translate the *intensity* of individual opinions. Yet in spite of these imperfections the vote remains the basis of participation in democratic societies.

There are other forms of social feedback. The market price of goods and services is the support for a kind of continuous vote represented by the vast numbers of transactions between buyer and seller. In making a purchase a consumer indicates a choice, just as he does in voting. The power of a boycott of certain products and voluntary restraints on buying during times of shortage illustrate the "macroscopic" effect of a mass of individual actions.

The stock exchange is another system of participation in real time. Each order to buy or sell is a sort of vote that modifies the market price and has consequences for the management of numerous businesses, for financial houses, and eventually for a multitude of workers.

Political leaders, labor leaders, industrialists, newspaper owners, producers of television programs, and directors of advertising agencies have long been trying to learn what people think, to anticipate their reactions, and to satisfy the needs and desires of the population. The suggestion boxes that industrialists place in company cafeterias, the letters to the editor in newspapers, the "open door" policies in big business, and the work of the ombudsman in European governments are limited but signifi-

---

like Yona Friedman, who say that all global communication is impossible because of the critical size of groups. We are moving toward a "poor world" fragmented into hundreds of little communities with communication between them reduced to a minimum.

cant attempts to obtain feedback at decision centers. These practices attempt to translate, at the highest level, the responses of citizens, consumers, and employees to the programs and measures that affect them.

But these rudimentary social feedback channels are a mockery compared to the power of the systems of descending information, particularly television and advertising. To speak of communication here, on the theory that the receiver "will get the message," is an abuse of language. There can be no true communication without the feedback of information and interaction with the source.

The Imbalance in Communications

Inundated with floods of descending information, citizens are condemned to playing the role of passive observer. The feeling of frustration that they experience results from the imbalance between the unquestionable educational effectiveness of communications systems and the weak efficiency of feedback channels that are supposed to allow everyone to express his opinion or to participate fully in the operation of the society in which he lives.

There is another imbalance, that between two new social classes: the "information rich" and the "information poor." The gap between them may increase with the utility costs of interactive networks in real time.

The explosive and uncontrolled proliferation of the media has created a condition of anarchy: a new form of pollution by information and a profound malaise on the part of all who must suffer the information without having the power to control it.

Today we are witnessing a reversal of attitudes as a result of the constant questioning and the pressures exerted by the younger generations. There exist powerful antibureaucratic and antielitist feelings among the students of many countries, along with the compulsion to criticize immediately all forms of excessive centralization of power. It is a secret war against the influence of what Ivan Illich calls the "radical monopolies": systems of education, health, news, entertainment, transportation, and organized leisure. It is a feeling expressed not only in meetings and the underground press but in a crowd of initiatives:

We are experiencing an increase in pressures brought by citizens on behalf of laws and regulations limiting the power of certain organizations by making them more open to the public. The rise of the press as a fourth power (after the executive, the legislative, and the judiciary) in the Watergate scandal is the sign of a firm will to reestablish the balance of powers and to prevent confidential information from being controlled and used for personal ends.

The fight for the safeguarding of privacy from all forms of electronic eavesdropping and the files in central data banks is another sign of the determination to readjust the balance of powers between those who hold the power to collect and store information and the citizens whose lives are recorded in the electronic files.

The investigations and publications of Ralph Nader and his "raiders" have shown the need for rigorous control of specialized government agencies that exercise monopoly power in some sectors of our daily life—such as health, education, food, and transportation. In the case of the Food and Drug Administration and the Federal Trade Commission, Nader's investigations have brought out the serious implications of decisions made in haste or under pressure from industrial groups.

The rise of consumerism and the expanded roles of consumer associations, parent organizations, neighborhood committees, and conservation groups are contributing to the increased power of social groups that play a significant part in the life of the country.

Student demonstrations and mass meetings, protest marches, and sit-ins that take place in front of television cameras are immediate forms of social feedback whose repercussions must not be underestimated.

In the United States the creation of Community Information Exchange Centers, located in small cities or in sections of large cities, help to bring people together on the most diverse subjects: mutual education, family and vocational guidance, drug addiction, hobbies, philosophical and religious studies, and conservation. These centers also operate as sorting stations for garbage, recyclable materials, and salvageable items.

Everyone knows the influence that Americans can exert through locally organized referendums on subjects of national interest. When questioned, they readily take stands on such subjects as the legalization of marijuana or abortion, educational reform, highway construction, urban renewal, and regional development.

These kinds of social pressures, added to the possibilities offered by the new interactive communications networks, will open up millions of channels of expression; little by little they will reverse and rebalance the flows of information at all levels of society.

The Media and Electronic Participation

The media have been quick to react to the rise of discontent resulting from the citizens' feelings of being left out of power. On their own initiative they have contributed liberally to the installation of new systems of social feedback. Apparently this was done first by radio and television networks, then by cable television companies, who were among the first to realize the social and commercial potential of an entirely new form of electronic mass participation.

The earliest instances of the expression of a collective response through radio and television have their own history. Some years ago the head of a large television network told the press about one such response. The engineers of the New York City Water Department were puzzled by the regular water consumption cycles that occurred every quarter of an hour and saw peak usage for short periods. On investigation they discovered that the cycles corresponded precisely with the times that advertising was broadcast by all the major television networks. Television viewers were using those few minutes of advertising time to get a drink of water or to visit the bathroom!

Two instances of polling in real time, carried out by French television several years ago, deserve mention. A team of television professionals decided to question the residents of Sarcelles, considered a model "bedroom town," on the problems of living in large suburban conglomerates. In order to get an instantaneous collective response, the directors set up their cameras one night on the heights surrounding the city, from which thousands of lighted windows were visible. They asked the viewers who were watching the program (about 70 percent of the inhabitants) to turn off their lights at the beginning of the broadcast and to turn them on again only if they wanted to reply in the affirmative to the questions they would be asked. The vision of thousands of lights coming on instantly in response to the questions of the television host excited everyone who participated in the event.

The idea had been taken up by television on the occasion of a public fund-raising campaign sponsored by the French Foundation for Medical Research. Everyone remembers the event. At a prime viewing hour, an announcer speaking for the foundation asked all Frenchmen to participate in a drive to benefit biomedical research by buying a "share of life." To estimate the number of viewers interested in the appeal, the announcer asked everyone who wanted to participate to turn off his television set for one minute. The drop in current registered by Electricity of France and transmitted to computers would indicate how many viewers had turned off their sets, and the result would be broadcast. The total was 3.8 million viewers; later they all went to their town and city halls to buy their "share of life." More than 20 million French francs were collected in a few hours.*

Radio stations in many countries allow listeners to call at the time of broadcast discussions. In the United States there are stations that devote almost all of their broadcast time to conversation with their listeners; this is person-to-person radio. In France radio and television programs that give listeners or viewers the opportunity to express themselves or to offer their help have enjoyed great success. Several years ago American and German television networks inaugurated "participation" broadcasts. Viewers responding to a news question that could be answered yes or no called telephone numbers designated for affirmative or negative answers. The calls were quickly counted and the results broadcast.

---

* This kind of public appeal is open to discussion and has had mixed reactions in France. I shall not pursue the controversy here; I only want to illustrate the potential for social feedback on a large scale.

More highly perfected systems of social feedback have been tried in the United States. One system uses terminals installed in homes; by pressing a button, individuals can register their opinions in polls. Another system uses survey checklists that appear in the daily and weekly newspapers. The forms contain boxes corresponding to the answers to the various questions. Readers check the appropriate box for each question, the survey forms are read and tabulated by computers, and the results are published in the next issue. Computer terminals have been installed in public places and in supermarkets so that consumers can inform manufacturers of their reactions to certain new products (Fig. 78).

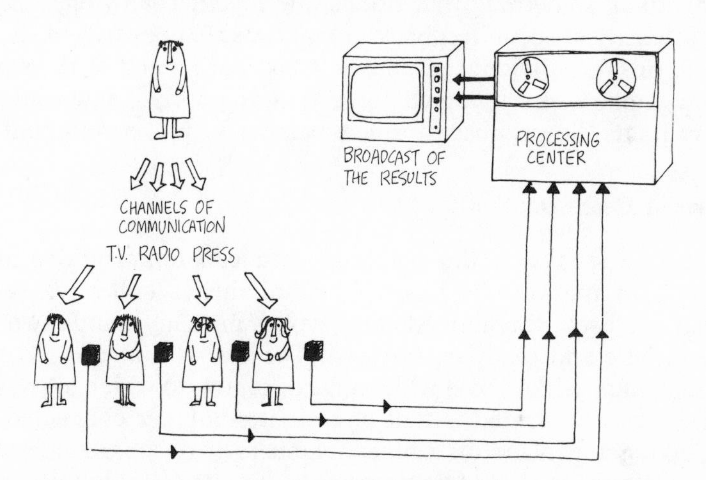

Fig. 78

Cable television companies are now experimenting with several kinds of interactive systems. Subscriber Response Systems (SRS) enable a single computer to collect information from the terminals of ten thousand subscribers in less than two seconds. In South Orange, New Jersey, four thousand cable television subscribers participated in a survey on programs and their quality. The Mitre Corporation conducted an experiment in Reston, Virginia, in which subscribers choose their own programs and communicate among themselves through individual "addresses" stored by the computer.

So far the press has not participated in social feedback operations on a large scale beyond the publication of opinion polls. The practice of publishing opinion polls goes back to the end of World War II; the surveys are a kind of social mirror that reflects for the nation a fixed image of its opinions and its choices on a wide range of subjects.

Experiments in social feedback in our day all stress one important point that has been confirmed by other research (notably that on educational systems using classrooms equipped to register a collective student response): collective feedback is valid for each participant only when he receives the overall results in real time. The students say there is a big difference between making a mistake along with 80 percent of the class and making it all alone. In any case, they want to know. Social feedback also seems to reinforce the willingness to participate; one wants to know more and to learn from the responses of others, as one is anxious about what others have said and how they reacted.

What is also striking in this kind of experiment is the sense of togetherness that connects and integrates the members of a group taking part in a large-scale investigation. Each person has the feeling of acting in a new dimension, of participating effectively in something larger than the individual, something that brings one together with one's equals.

The intensity of social feedback in real time that consists of thousands, even millions of individual responses is fascinating and at the same time disturbing, like some untamed force that is poorly known and poorly used yet holds the promise of a new balance of power and control.

Problems of Representation

When one speaks of the potential of telecommunications and data processing in the various forms of simultaneous collective responses, or social feedback, this immediately evokes in many minds two images, both futuristic and easily caricatured. The first is that of a "continuing electronic referendum" on a wide range of topics, to which citizens would be subjected. The second is that of a giant computer connected to each voter, taking the place of cabinet members and congressmen in their roles as planners and coordinators of the country's economic and social life.

The two possibilities are as absurd as they are unlikely. Such systems, to be at all effective, must assume that citizens are informed to such a degree about the problems on which they have to form opinions that in fact they would have to spend so much of their time collecting, organizing, and studying information that they would have no time for other activities.

Fortunately the computer is not ready to make every living room a center of government. Moreover, this kind of continuing referendum on a nationwide basis, even if it were possible, would be extremely dangerous. The immediate response of millions of citizens to the questions that a president might ask them directly on the little screen would cause a form of short circuit, resulting in an enormous loss of energy. Information rising instantly from the entire base of the social pyramid to its summit would have the paralyzing effect of a social electrocution.

More than ever we must take into account the necessary delays in response times that are a part of social systems—the hierarchy of levels that allows intermediary bodies and representative organisms to act as transmission lines. The absence of friction, delays, and restraints can lead to extremely dangerous, self-amplified oscillations, as studies of servomechanisms have shown. Filtering, buffer effects, and even disorder, introduced by interactions between individuals, protect the social system and allow it time to adapt to rapid change and new situations. The delaying factors also alleviate the volume of responses, eliminate "noise," and in the long run draw out significant information and tendencies.

To be effective, a participative system must take into account both the response of the people and the intensity of that response. Without the dimension that intensity gives it, a response is empty; one soon realizes how difficult it is to moderate the result of the affective or impassioned reactions of a mass of individuals who are poorly informed about the situation on which they are consulted. This is especially true in evaluating the intensity of the responses of minority groups.

Instead of national electronic referendums, we need decentralized systems of participation that permit continuing control and planning of social and economic activities at the local level (neighborhood, business, city, state, or region). The human organism and biological systems in general offer numerous models for the decentralized regulation of equilibriums. Such decentralization deals directly with the function of representation.

The representative (local elected official, congressman, union officer, administrator) does not need all the available information on a given subject. He cannot be at the same time a public opinion institute, a storage bank, and a transmitter-receiver that decodes and transmits faithfully the messages from his constituents. The representative can select, distort, amplify, or hide information to serve his personal ends. More than the perfect transmitter of information, he is the creator of a new form of information, a principal actor in the greater participative system.

Out of this subtle game of transaction, filtering, and negotiation, the function of representation is born, emerging at the "macroscopic" level, as does every systemic property. The question is not one of knowing whether to get rid of representatives considered out of date in the technological plan for the communications revolution; it is a question of knowing how best to use the interactive participation systems, electronic or not, at all levels of organization, in order to strengthen the function of representation and especially to restore the balance of powers among representatives, the represented, and the managers.

Still we are not sure which of the electronic means could effectively help representative bodies, pressure groups, lobbyists, labor unions, consumer associations, employee committees, and municipal councils. We might try to strengthen the role of the representatives by giving them access to an impartial and objective "service of experts" made up of electronic interactive participation systems linked to citizen groups.

The interactive participation systems created for special purposes will play increasingly important roles in local government, international organizations, and large symposiums. Several countries are already experimenting with continuing referendum installations that permit qualified responses in the place of the simple yes or no.*

Such systems will be used first at the local level, then extended through a series of interconnections with all the major professional fields and finally with entire geographic regions. The initial impact of these systems will probably be felt first in the business world.

Contrary to what classic theories of management advise, more and more attention must be given to the flows of information that rise from the base of the pyramid toward the decision centers. No one can appreciate a problem better than the person who is closest to it. In the United States and in Europe general managers compliment themselves on being able to make fast decisions—but how long does it take to apply the decisions that come "from above"? In Japan arriving at a decision is a slow process because everyone participates, but once the decision is made it is put into effect almost immediately.

Advantages and Dangers of Society in Real Time

Instant access to information and the use of electronic systems for participation in real time hold out great hope for a transition to a more just and more humane society. At the same time they represent one of the most serious threats humanity has ever faced: the risk of concentrating power in a few hands has never been so high. Yet the opportunities for bringing citizens closer together have never been so great.

The redistribution of power that data banks allow permits a more conscious participation of individuals in the general functioning of society, in its major decisions, and in the regulation of its equilibriums. The social feedback loop, which we perceive now at the level of observation of the macroscope, could in the very long run be one of the predominant elements in the regulation of the metabolism of society. This feedback loop will contribute to the control of energy consumption and the growth rate, the adjustment of production to needs, and the control of the production of wastes and the cycles of recovery and recycling.

During the worst of the energy crisis, public opinion was impressed by the breadth of the results—within a few weeks and on a national scale—that followed restrictions on travel and the regulation of thermostats. Through the feedback of results, everyone discovered the power of collective efforts coordinated and synchronized toward a specific goal.

---

* More examples are given in the sixth chapter, on education.

The big difference, compared to well-known movements in history (wars, fascist and totalitarian dictatorships), was that for the first time such movements could be coordinated by the citizens themselves, in their own interest. Social feedback makes it possible to respond to demand and to need—to adapt to an environment experiencing rapid evolution, to anticipate and use events as evolutionary factors instead of managing successive crises. The large newspaper organizations and the television networks will be able to tailor their publications and their programs more effectively, satisfying the public's aspirations while continuing—by maintaining a dialogue with the public—to raise its general level of knowledge.

But democracy in real time offers not only advantages; badly directed and controlled, it can lead to the worst of dictatorships. In fact a more sensitive, more interactive society that depends on complex regulation systems becomes still more vulnerable to destruction and to distortions of all kinds. It is like any other complex living organism. What guarantees can one offer the public to assure it that the interactive networks will not serve the interests of small political or business groups rather than those of the public? It is easy to falsify or manipulate the data that result from opinion polls in real time, through the selection of criteria that modify their treatment by the computers and the posting of the results.

Installation costs for electronic systems that provide instant access to information might be so high that only a few large industrial firms would have the means for developing them, using them, or controlling them in their own interests. At whatever level it occurs, social feedback clearly has value and interest only when it comes from *all* the individuals concerned. Can you imagine a situation in which the cost of terminals was very high or there were long delays in installation and some citizens were deprived of their right to vote while others, in more comfortable circumstances, were favored? Once again decentralized participation at several hierarchical levels is necessary within the framework of a public service from which all citizens can benefit.

The dangers of manipulating flows of information being fed back to the decision centers and the dangers of invading citizens' privacy by building up data banks on them are obvious. An information network linked to computers and continuously interrogating the terminals in specific homes in order to learn those people's questions or tastes could become the basis of a gigantic electronic file on the individuals. The shadow of Big Brother described by George Orwell in *1984* stands out as data are increasingly centralized through electronics.

Political problems posed by social feedback have been very little studied so far. I have spoken of the role of representatives; on another level, how can one establish continuing citizen control of the groups or bodies that have charge of the programming of computers and the maintenance of networks? How can there be control over the way in which questions are asked?—a particularly delicate problem when one understands the influence of the wording of a question on the persons chosen to answer it. How can one protect oneself against momentary "gut" reactions? How should one treat the necessary maturation time and the delays inherent in social systems? We do not know very well the response times of these systems. The cumulative effects of a series of seemingly insignificant stimuli, taken up by the media, are capable of creating a climate of tension or collective hysteria. An electronic participation system could amplify such reactions through positive feedback and lead to collective behavior that would be catastrophic.

An entire science of the dynamics of complex social systems remains to be established. Shall we be successful in respecting our individual liberties as we install the cybernetic mechanisms of regulation in real time that are so grievously wanting in our social systems even as they form the basis of biological systems?

Five

Time and Evolution

Everything is linked to time, even the full meaning of words. Any vision of nature and society that wants to be comprehensive cannot ignore the vast problem of time; it determines even our manner of thinking.

The contrast between *physical time,* a frame of reference that is outside events and phenomena, and *psychological time,* which is rich with the intensity of living experience, reveals itself in everyday language as well as in the languages of organization and data processing. We speak of time gained or lost, of shared time and real time, of free time and the lack of time.

Beyond the difference between physical and psychological time lies a fundamental question: Do not many of our understandings and irreconcilable points of view arise from the use of strongly "polarized" concepts through implicit reference to a privileged direction of the flow of time? These concepts have an entirely different emotional meaning, depending on whether the unconscious reference is to time that aims toward entropy or toward organization—according to a causal explanation ("pushed" by the past) or a final explanation ("pulled" by the future). Does this also explain the unreconcilable conflicts—between determinists and finalists, for example, or between materialists and spiritualists—that spring up as soon as the discussion turns to evolution?

To go beyond such conflicts, we must free ourselves from what I call our *chronocentrism.* The term may seem a bit strange; I use it here in relation to two better-known terms, geocentrism and anthropocentrism.

Thanks to the theories of Copernicus and Galileo we have succeeded in getting rid of our geocentrism, the stifling idea that the earth is the center of our world. It was just as difficult to escape anthropocentrism, which put us at the center of all living things. Thanks to the theory of evolution, man is again one species among thousands.

Yet the most difficult threshold remains to be crossed. We are prisoners of time and words. Our logic, our reasoning, our models, our representations of the world are hopelessly colored by chronocentrism (as they formerly were by geocentrism and anthropocentrism). From chronocentrism come the conflicts that paralyze our thinking. Can we free ourselves from them?

It is difficult and dangerous to tackle the concept of time. Each of us feels deep inside that he must struggle fiercely, step by step, to preserve the concept, to continue to let himself be guided by this vital thread to which we cling as though it held our universe together. To break the thread would be to risk undoing, stitch by stitch, the net woven by preceding generations, the web in which our past is imprinted and our future constructed.

Nevertheless we must pull gently on this thread to see where it leads and to learn whether it forms a closed loop. To study the world through the macroscope is to try to perceive, beyond details, the great principles that tie us to the universe. Without the attempt to leave the tunnel that time has drawn us into, there can be no constructive dialogue between the objective and the subjective, between observation and action.

1. KNOWLEDGE OF TIME

Through our sensations we project on the universe the "reality" of *terra firma,* of geometric space, of time that never stops. Most of the major laws of physics come from the interpretation of information communicated directly or indirectly by the eye and the muscle and then stored in the memory.

The eye is an instrument that is particularly well adapted to recognizing forms, detecting changes, and perceiving movement. Man's muscle allows him to measure and compare weights and efforts; it leads him to interpret his relations with the outside world in terms of *forces.* Memory accumulates and concentrates time, whose course is inscribed in the web of our consciousness.

We are accustomed to describing events by using four coordinates: the three spatial coordinates (*where* the event occurred) and the coordinate of time (*when* it occurred). Just as it seems to us impossible to conceive of the outside world without relying on geometric properties, so are we unable to describe it without referring to the passage of time. But where does the idea of *before* and *after* come from?

Memory and expectation point past time (the before) toward the future (the after). The two modes of conscious behavior are perceived to be different and asymmetrical. We know that we can act on the future but not on the past. We are conscious of *knowing* the past to the smallest detail, while the future seems to be enveloped in the uncertainty of chance and the possible (Fig. 79).

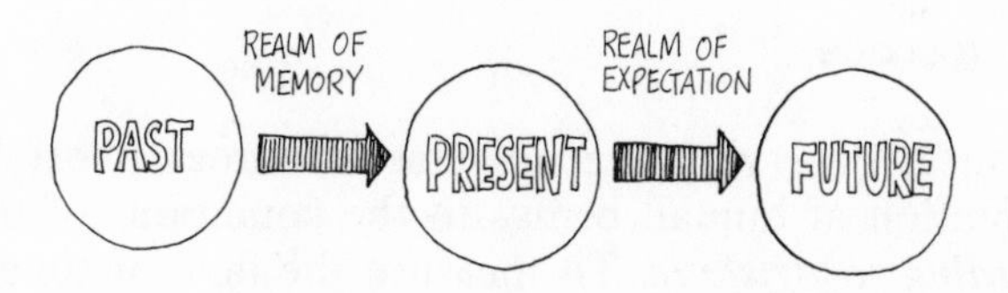

Fig. 79

When we stop the machine tape on which the movement of a pendulum is being traced, we see only a continuous line; when the tape is started again, the line becomes sinusoidal. For the pendulum of a clock there is no process of time. It is our consciousness that creates duration and, like the machine tape, records past information as a series of peaks that we can number. By deciding that one is before and another after, our consciousness can develop a chronology of events.

At the same time as the concepts of force, movement, and before and after, there appeared two concepts that are mutually irreducible, continuity and discontinuity.

We have the sensation of continuous movement as we follow the trajectory of a moving object, as we watch the road unwind under the wheels of a car, as we contemplate a liquid flowing without interruption. But if we turn our attention even for a moment to the location of the moving object, to one stone in the road, or to one drop of the liquid, our concentration on discontinuity immediately destroys the sensation of movement. One cannot concentrate on continuity and discontinuity at the same time.

In the same way, the flow of time can be seen either as *duration* or as a succession of *instants*. Intelligence is accustomed to cutting up continuity into moments and objects of determinate shape. Contrary to intuition, which according to Bergson is the feeling of things in motion, intelligence freezes what it isolates from the flow of time. Since its method is analytical, intelligence can understand movements or flows only as a succession of juxtaposed still positions.

This limitation on our perception of nature has great significance. It is found at the root of the distinctions between flow variables and state variables (see p. 73) and between the ondulatory and corpuscular aspects of a fundamental particle. It was in order to overcome such dichotomies that the concept of complementarity was introduced: each entity in nature can be conceived at the same time in its continuous aspect and in its discontinuous aspect.

Time in the Evolution of Thought

A short history of the various conceptions of time in scientific and philosophical thought will help us to sort out the paths of the contemporary theories. Is the concept of time an objective idea, independent of our consciousness observing the universe? Or does it originate in the rigorous adaptation of human beings to the conditions of the universe?

Time according to Aristotle. To measure the flow of time one relates it to space through movement. For Aristotle "time is the quantity of movement." Thus one divides space into as many gradations as can be linked together, either by the movement of a shadow on a sundial or, later, by the movement of the hands of a clock. In the same way a road can be divided into segments of equal length, identified by markers and linked by the movement of a vehicle—which brings us again to the measurement of time by the regular speed of a moving object.

Time according to Newton. Newton identified himself with the search for an "objective" time that was outside phenomena, a flow of time that would run through the universe of its own accord. In laying down the concept of *universal time* as the basis of his mechanics, Newton was led inevitably to the principle of *absolute space,* according to which each place or each position is identical in every respect to any other in the universe. For Newton there must be privileged axes of reference that are absolutely immobile and that make it possible to describe the universe and the processes that occur in it.*

The irreversible time of Carnot and Clausius. The thermodynamics that sprang from the works of Carnot (1824) and Clausius (1865) no longer calls expressly on the concept of space but on the concept of time. It speaks now of transformation and no longer of movement. Irreversibility does not exist at the microscopic level, in the simple, homogeneous systems that are the concern of classical physics. Physical laws obviously take into account the passage of time, but not its sign; negative time and positive time play the same role. If t were changed to $-t$, the world would be a strange place, but there would be no fundamental conflict with the laws of nature. It is only when the phenomena of dissipation, diffusion, friction, disorganization, transfer of energy, and especially complex systems are considered at the macroscopic level that the irreversibility of time enters the picture (see p. 103).

What can we infer from this? That all systems that are sensitive to the passage of time have in common the ability to move from a state of high organization to a state of disorganization, or a state of higher probability. Thus it is only in complex systems that time seems to run irreversibly and toward increasing entropy. The arrow of time and the arrow of entropy point in the same direction.

A statistical clock has just been added to the moving clocks of Aristotle and Newton, and this clock plainly indicates time as *irreversible.*

---

* The Newtonian illusion of a time of things, an absolute time, was vigorously defended by Samuel Clarke in his famous correspondence in 1715 with Leibnitz, who considered time an "order of events." Later, with Kant, time moves to the side of the subject by becoming "an *a priori* form of sensitiveness" (*Transcendental Esthetics,* 1781).

The time of Einstein. The theory of relativity introduced a new upheaval, the transformation of space into time, or the "spatialization" of time (time and space being equivalents). Henceforth we can speak only of a "space-time continuum." For relativists time does not "pass" and matter is unfolded in both its "temporal thickness" and its "spatial span"—which means that time, like space, is an actual span. We can no longer refer to a "universal time" and an "absolute space." The properties of space-time depend on the speed at which a moving object travels, and at speeds approaching the speed of light, space-time "contracts" around the moving object. But the time of relativity, like that of classical physics, remains reversible.

Time according to Bergson and Teilhard. Bergson and Teilhard place the direction of evolution over that of entropy. According to Bergson, "all our analyses teach us that life is an effort to climb the slope that matter descends." Teilhard measures the duration of evolution by the series of transformations that lead matter, life, and society toward states of higher complexity. "We are already prepared to observe that life, taken in its entirety, manifests itself as a current opposed to entropy. . . . Life, contrary to the leveling play of entropy, is the methodical construction of an organization that ceaselessly grows bigger in the most improbable way." For Teilhard space-time takes the shape of a cone: the point of the cone is the outcome of cosmogenesis; God is Omega, the end.

Thus the distinction between the two great currents of evolution and entropy is clear in the minds of these two authors. One "climbs" toward life and the mind; the other "descends" toward matter and multiplicity. The "ascent" of life seems to have to be measured by a thermodynamic "clock" whose hands turn in a direction opposed to that of the clock of Carnot and Clausius, for instead of entropy it is complexity that appears to increase locally.

Bergson introduces another fundamental asymmetry, that between the time of invention (creative duration) and the time—almost instantaneous—of reproduction.

The duration of the universe goes hand in glove with the "possibility of creation that can take place there." Since every determinist process is foreseeable, reversible, and reproducible, the freedom of the creative act renders this act unforeseeable, irreversible, and impossible to reproduce. In the creative transition from the virtual to the actual—or, as Aristotle said in an especially illuminating way, from the "power" to the "act"—there are unlimited possibilities. The realization of just one among them immediately excludes all the others. This is what gives a work of art its unique character and its value; the moment of creation is an "historic" moment, the moment of copying is commonplace. That is why the future is not given alongside the present; creation requires duration.

Time in Contemporary Theories

In a stimulating book published in 1963, *The Second Principle of the Science of Time,* a French physicist, O. Costa de Beauregard, provides the first elements that make it possible to reconcile the reversible time of relativity and the irreversible time of the consciousness. He suggests a fruitful hypothesis concerning the manner in which consciousness meshes itself in the universe through the dialectical process of observation and action. Thus the hypothesis integrates the ideas of thermodynamics, information theory, and relativity.

Costa begins with the work of Szilard and Brillouin leading to the equivalence of negative entropy (neguentropy) and information, that is, to Carnot's principle generalized (see p. 134). Its main conclusions deserve recall. Information—which is order, organization, and improbability—is the opposite of entropy—which is disorder, disorganization, and probability. Entropy measures the lack of information in a system. Information is thus the equivalent of negative entropy. Every experiment, measurement, and acquisition of information by a mind consumes negative entropy. Thus a tax must be paid to the universe, and that tax is the irreversible increase of entropy.

Yet the mind can create negative entropy, thereby increasing organization, order, and the quantity of information in the system in which it is found. The global system remains subject to the law of universal degradation.

Carnot's principle generalized fails to answer satisfactorily three questions: Why does the inquiring consciousness explore the universe only in the direction that sees an increase in entropy, that is, the direction we call "time"? What is the actual difference between neguentropy and information? Why are we conscious of such an asymmetry between observation and action (the first "costing" less than the second)—or why is it easier to destroy and copy than it is to construct and create?

For Costa the direction in which every inquiring mind explores the universe is adaptive. As soon as an animal or a man opens his eyes on the world around him, information from outside is linked to an inward flow. Information appears in the form of waves sent by a radiating source—light, heat, sound. Living beings adapt little by little to the direction of the waves from these sources. This adaptation becomes a rigorous condition of survival, since living beings can act on their environment only to the degree that they receive and intercept information coming from it.

But man can observe phenomena only in the direction of disorganization, since every acquisition of information is paid for by the increase in entropy. Thus each observer follows the course of time by "accompanying" the phenomena he observes. Living is an arrow pointed toward dying; without this imperative condition we could not observe phenomena. And without information all creation would be impossible.

An answer to the first question may now be attempted. It is not matter that advances by "evolving" in a static space-time framework. If anything advances in the spatial-temporal block, it is the inquiring consciousness. The universe is spread out over its entire temporal dimension. Time is given, it does not pass. But because of its adaptation to the conditions of the universe, the consciousness, in order to acquire information, can explore it only in the direction of increasing entropy (the direction of time). The observing consciousness meshes itself in the universe like a funicular on a one-way trip.

On the other hand, by creating new information the consciousness accumulates something in an "opposite" direction—in another dimension, that of creative duration aiming toward ever higher levels of complexity.

The second question concerns the difference between information and neguentropy. Costa de Beauregard cannot avoid reintroducing the subject in the world of objects: every creative or inquiring mind has its influence on the increase of entropy in the universe. Must one dare to use this bridge between the subjective and objective worlds? If one crosses the gap, neguentropy appears as the objective counterpart of information.

We have seen that all information can be measured in a quantitative way (in bits, for example) and that to accomplish this measurement the meaning of the information must be disregarded. Neguentropy is completely neutral and objective. It travels in a telephone cable or in a computer, but it enters and leaves in the form of meaningful information. For the consciousness each item of information possesses a different sense, meaning, and subjective value. The mind distinguishes without difficulty between information of high value and information of no interest, even though both amounts of information may be measured by the same number of bits.

Are information and neguentropy perhaps subjective and objective aspects of the same form of potential energy? Costa de Beauregard does not answer the question definitively. However, the transition from one form into another, through observation or action, does imply two asymmetrical processes that strongly suggest the transition from the subjective to the objective.

Classical determinism regards free action as being "impossible" on the scientific level (theory of the epiphenomenal consciousness). Observation, however, raises no difficulty. This is because the consciousness has two fundamental modes of activity (see pp. 131, 139). One corresponds to the transformation of neguentropy into information. This is the process of observation, where information means the acquisition of knowledge (Fig. 80).

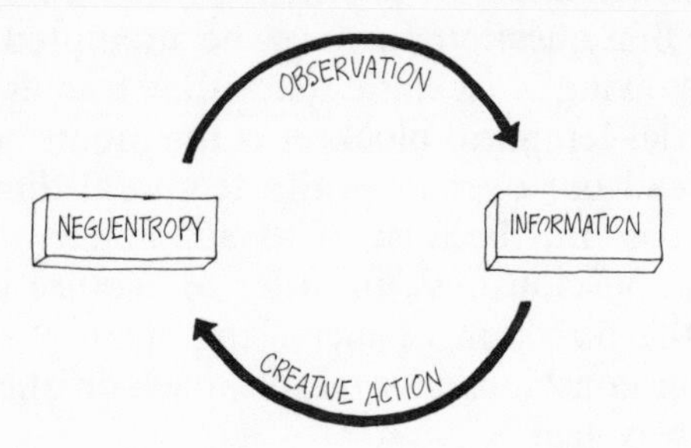

Fig. 80

The other corresponds to the reverse transformation, of information into neguentropy. This is the process of action and creation, where information means the power of organization (to give form to something). In one case the mind is informed, in the other it informs.

The first process actualizes, or puts to use, the information that has been acquired—in distribution, reproduction, and copying. This process costs little in neguentropy (in preexistent potential energy). That is probably why measurement and observation always seem to raise no difficulty.

In contrast, the reverse process of creative action costs very much in information. That is why the creation of an original (as opposed to making a copy) seems so difficult. The popular expressions "actions speak louder than words" and "easier said than done" also reflect this principle.

The temporal difference between the two modes of activity of the consciousness is also very important. The time of actualization can be instantaneous, as Bergson foresaw; it depends only on the efficiency of the duplication and broadcast media. In contrast, the time of free action and organization is related to creative duration. The time of actualization is time that "spreads out," the time of ontogeny, of our physical life. Opposed to it is time that "adds on," the time of phylogeny, of evolution, of creative duration.

Contemporary theories suggest that the conventional direction of the passage of time, measured by the passing years (and headed from the past toward the future), is the result of an adaptation of the consciousness to the conditions of the universe.

The time in which an observation occurs is certainly headed in the direction of increasing entropy, in accord with the direction of conventional time. What about the time of creative action? It seems to belong to a time that is qualitatively different, apparently reversed by the consciousness and pointing in a direction opposite to entropy—the direction of increasing complexity. How can we distinguish this direction from that of conventional physical time?

The chronocentric attitude is uncompromising; it refuses to consider the complementarity of two "qualities" of time—just as physicists once considered only evolutions that pointed toward an increasing entropy and refused to integrate into their theories the possibility of a biological evolution.

Chronocentrism adopts a logic of exclusion; it accepts *only* causal explanations—and emphasizes, in this case, the principle of sufficient reason and the assumption of objectivity. Or, on the other hand, it accepts *only* final explanations arising from some "act of faith" and subjective action.

The significant difference between the two extreme attitudes is that causal explanation is strongly emphasized in our education and our culture. Causal explanation is based on experiment, demonstration, and scientific proof, while explanation by finality allows neither irrefutable demonstration nor scientific proof.

One advances into so delicate a realm only with a certain caution, proceeding by successive stages: demonstrating first why our logic is led astray by circular causality; illustrating then the obstructions that result from the adoption of either causality or finality as the one method for explaining phenomena; and proposing, finally, a new route that may make it possible to overcome these conflicts.

2. THE PRISON OF TIME

The Link Between Chronology and Causality

The cybernetic feedback loop has many interesting properties, some of which are linked to time and have not yet been mentioned. Having discovered them, the first cyberneticians were obliged to introduce finality, or purpose, into the world of machines.

In an information/decision/action loop, information on the results of past actions is the basis for the decisions that will correct a present or future action. Because decisions are made to achieve an end, the consequent action is purposeful; such a loop illustrates the occurrence of an intelligent act (Fig. 81).

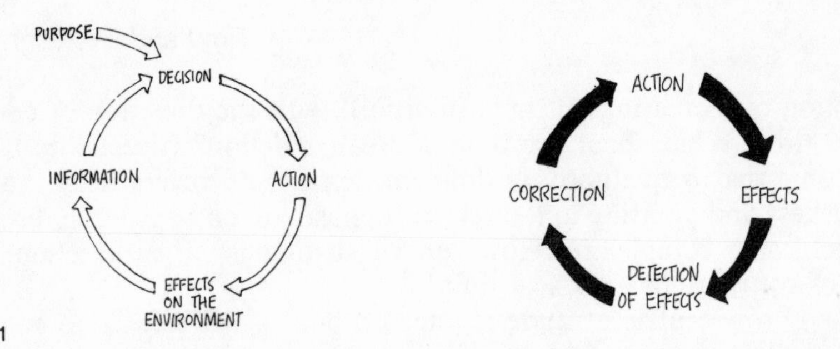

Fig. 81

Not only men achieve intelligent acts; there are also the cybernetic machines, the servomechanisms. Their "decision" mechanism is also embedded in a feedback loop. The character of this loop is very general, and I have already given numerous examples of it.

Consider the general circuit of any feedback loop and then ask, does cause precede effect or does effect precede cause? It is impossible to say; there appears to be no distinction between them, and they cannot be split apart in time. Causality follows the entire circuit of the loop; so does finality (Fig. 82).

Fig. 82

We are forced then to speak of circular causality as opposed to linear causality, which is represented by a vector superimposed on the time axis, where cause coincides with "before" and effect with "after." Thus a feedback loop is like a serpent that bites its own tail.

The loop of circular causality must not be confused with a cycle. A cycle is always subject to unidirectional time; it is an infinite repetition of the same sequence of events. There is no "becoming" in a cycle, and a cyclical succession can be measured by any clock. However, in a circular causality loop *the arrow of time appears to close on itself.* It cannot really be said that time "passes"; it is balanced by something else, a kind of conservation of time.

As soon as the chronology of events is questioned, our logic loses its footing and seems ill at ease. Why? Because only chronology permits explanation by causes. To be forced to abandon the principle of causality even for a moment profoundly shocks our logic. We may find it amusing to watch a film being run backwards. But our logic is completely disarmed in the presence of a "vicious circle"; because of the circulation of causality, we no longer know "by which end" to take hold of things. Thus there is a close analogy between vicious circles and feedback loops.

We seem to be caught in a vicious circle whenever we look for the origins of a complex system—as in the familiar problem (which came first?) of the chicken and the egg. Or that of the origin of man: every man or woman is born of a couple, each of whom was born of another couple, and so on. To break the circle it was necessary to conceive of the origin of humanity in a "first couple" created by divine will. The same is true for the origin of life: life depends on a very small number of basic organic compounds that are believed to have been made exclusively by living systems. How could life have begun in the absence of these substances? The answer is that the first cell was created by God, or—same thing—that the first cell appeared abruptly, fully assembled, entirely by chance.

What does reason do to get rid of the irritating logical problem posed by a vicious circle? *It opens the circle.* It cuts the circle at an arbitrary point, which allows the circle to stretch out straight along the conventional arrow of time. At the same time it recovers the familiar relationship of before-and-after between cause and effect (Fig. 83).

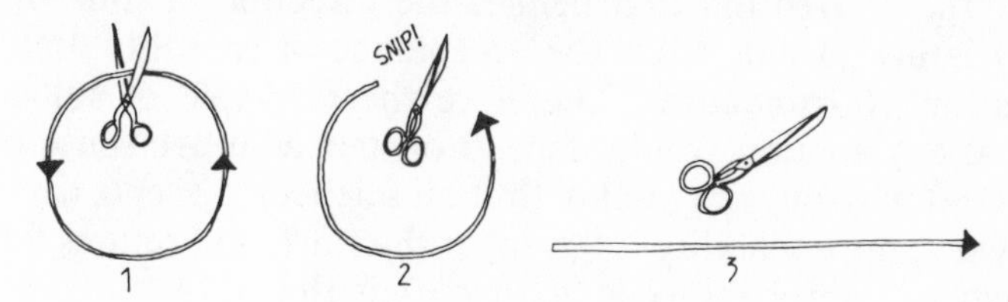

Fig. 83

This artificial cut into reality will have important consequences:

Causality appears to be the only method of explanation; we are forced to return, cause by cause, toward a "first cause" lying in the past.

Time "passes" again, for explanation by causes belongs to the process of observation, which points inevitably in the direction of increasing entropy.

We are obliged to adopt a reductionist approach.

In opening the circle *even with the slightest cut,* we allow an aspect of the whole to escape forever. Now complementarity makes room for a certainty that is limited to a single aspect of reality (Fig. 84).

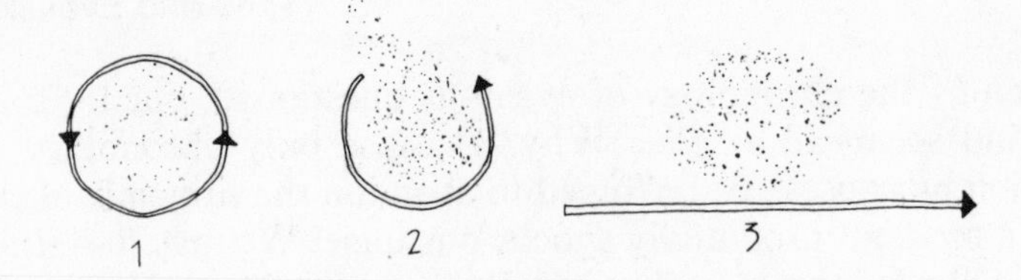

Fig. 84

This is what happens in every analytical approach. Incapable of considering all the interdependencies of the functioning mechanisms of the brain or the cell, we isolate several loops that seem to be essential, and we open them in order to find the familiar relationship of cause and effect. In this way we can explain perfectly well certain aspects of cerebral or cellular behavior through molecular reactions. And we will probably come to do this better as time goes by. But we know that something is escaping direct observation. Is is life? consciousness? the "soul"? I reject any vitalism that postulates the existence, in the heart of living matter, of a principle that will forever escape scientific knowledge. I say simply that the principle of sufficient reason or causal explanation reveal only one aspect of reality, owing to profound limitations linked to our perception of time.

Why are there such limitations on our reading of the phenomena of the universe? Probably because of custom that originates in the adaptive psychological meaning of "before" and "after." This custom makes a succession of events appear logical to us only insofar as there is *chronology,* insofar as the arrow of time points toward increasing entropy. Without knowing why, we have associated *chronology* and *causality.* The result is that "the convention that defines the direction of time by increasing entropy is inseparable from the acceptance of causality as a method of explanation" (Grunbaum). Therefore the principle of sufficient reason or causal explanation would depend on our adaptive sense of time. We understand why physics (and with it all science) "accepts causal explanations, where improbability is 'given' at the start, and refuses final explanations, where improbability is 'gathered' at the end."

Irreducible Points of View

The limitations of our thinking reach their bounds when we consider the phenomenon of evolution in its entirety, from the formation of living matter to the appearance on earth of living systems and social systems. The discovery of the great history of life and of man was made in reverse—from the complex to the simple, from subject to object—in accord with the "entropic" direction of observation in search of causes—that is, toward the past.

Man is breaking open one by one the vicious circles of the origins that have imprisoned his thinking. The circle of the origin of man is opened: the theory of biological evolution shows that man descends from simpler organisms that preceded him. The circle of the origin of life is opened: the first cell is the result of prebiological evolution. The circle of the abiogenetic appearance of organic substances is opened: they were formed in the course of the geochemical evolution of the planet.

When each of the circles is "stretched out" and the vectors are pointed in the conventional direction of time and then placed end to end, they reconstitute the greater vector of the evolution of matter, life, and society in that part of the universe that is our planet (Fig. 85).

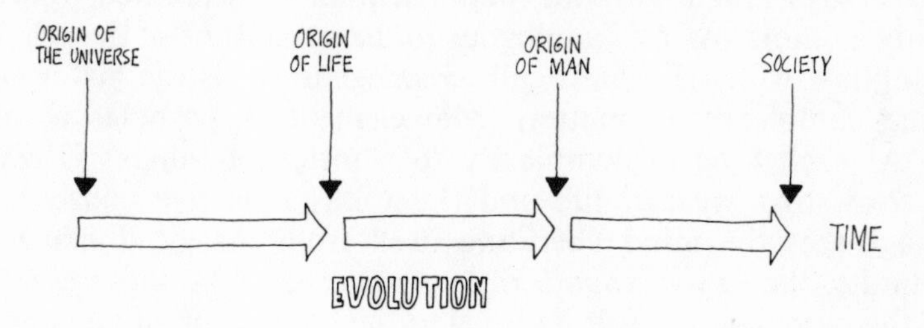

Fig. 85

Here we find one of the principal theses defended in this chapter: that the conventional direction attributed to this generalized vector of evolution leads to irreducible points of view.

The conflict between materialists and spiritualists can be traced back to modes of thinking and the use of expressions closely related to the acceptance of a conventional direction of the flow of time. According to the materialists, matter was present before the mind; according to the spiritualists, the mind existed before matter. Thus there came about a kind of hierarchy of preexistence, with greater value assigned automatically to what was there "before." This polarization is recognized in the expressions "initial impulse" and "final attraction"—matter being pushed (explanation by causes) or pulled (explanation by ends). How could the future both manifest itself in the present and be the cause of it? On the contrary, it is obvious that the past determines the future.

There is the same type of conflict between the Darwinists and the Lamarckists, or more generally between determinists and finalists; the struggle is fierce and the lack of understanding often complete. For the former, to admit any influence of the environment on heredity is to open the door to the spectre of a design of nature willed by a supernatural entity. For the latter, on the other hand, to think that molecular reactions occurring at random can condition heredity and the perfection of an eye, or that they determine thought and behavior, is to reduce what is most "noble" in life to mere matter—and thus to inferiority.

These two kinds of attitudes, carried here to their extremes, are shared by a large number of scientists and philosophers throughout the world. They illustrate a debate that it is helpful to personalize because it poses clearly the problem that interests us here. To do that, we might refer to two French authors whose works have stirred up a controversy that has not yet been calmed. The controversy was caused by Pierre Teilhard de Chardin's *The Phenomenon of Man* and Jacques Monod's *Chance and Necessity.*

Teilhard says that mind and matter cannot be separated, that there exists only a "mind-matter" analogous to the space-time of the relativists. All evolution—which Teilhard calls cosmogenesis—is the history of the increasing complexity of matter, from elementary particles to human society. At each level of complexity the "inside of things" is revealed in properties that we call life and (later on) reflective consciousness. Each stage sees the mind liberating itself from matter. Pushed to its logical limits, the *law of complexity consciousness* (the more complex a system, the more conscious it is) leads to the integration of all consciousnesses in a single God, the point of convergence of all evolution.

Monod says that there is no one concerted evolution of the universe but many evolutions that can be studied at the level of biological systems or at the level of social systems. In biological systems evolution is the result of chance mutations that cause changes in the genetic heritage. These changes are retained from generation to generation; this is the property of *reproductive invariability.* The environment acts as a filter, keeping only the best-adapted species. Life and thought are emergent properties, explained by the play of molecular interactions. The illusion of the "design" of nature is based on the teleonomic (from the Greek *teleos,* far, and *nomos,* rule) properties of complex systems, particularly enzymes, whose behavior seems to be directed toward a goal. Biological evolution is the result of the play of invariability and teleonomy.

In my eyes Monod and Teilhard are both right—Monod right in defending the principle of objectivity, Teilhard right in searching for a meaning in evolution. But both are probably wrong in using the approach and the language of the other side. One important point clearly deserves clarification. To do that, we must analyze the causal explanation, to which Monod refers implicitly, and the final explanation, which serves as the basis of Teilhard's system. Then we shall compare the two approaches, seeking ways to go beyond this choice.

The Causal Explanation: Divergence

Our science and our philosophy are founded on observation. They rely on reason (the principle of sufficient reason), objectivity (the premise of objectivity), demonstration, scientific proof, and the reproducibility of results. We can be rationally certain only after having explained by causality (the same causes produce the same results), verified, and demonstrated the validity of our theories. This is the rule for all good science.

However, as the works of Grunbaum, Reichenbach, and Costa de Beauregard suggest, the principle of sufficient reason, like that of causality, comes directly from our adaptive direction of time. Phenomena are significant for science (and observable) only when they occur in the direction toward which the life of those who observe them is also flowing. We would then be reduced to being absolutely certain only of what is decomposing and able to demonstrate perfectly only what is being destroyed. We would understand much better how things become disorganized than how they become organized.

Because of this, science goes spontaneously toward the past, toward the origins, to seek certainty. Every cause can be linked to a previous and more general cause. Having left the top of the tree, we descend toward the huge limbs that branch out from the trunk. From the millions of people on earth, we come to the "first couple"; from the abundance of forms of life, to the "first cell"; from all matter present in the universe, to the "primitive atom." By pushing causal reasoning to its limit, we must come to cosmological explanations of the type in which all neguentropy, all improbability is *given* to begin with. Out of this primary sphere of energy the universe begins to expand, entropy increases, and time passes (Fig. 86).

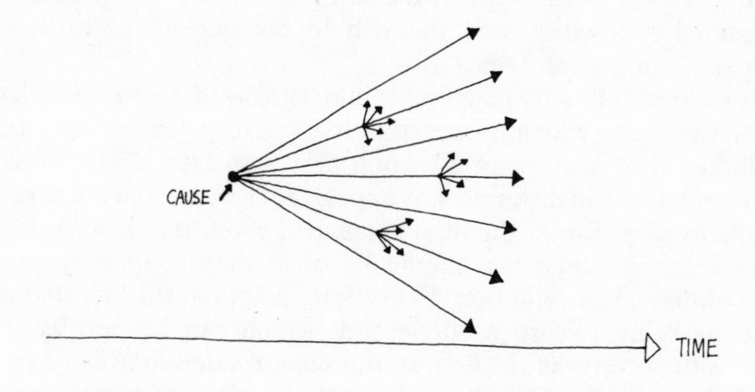

Fig. 86

Retraced in the opposite direction from a point in the past, all evolution founded on causal explanation can only be *divergent;* we see only arborescences like the tree of evolution or any family tree. From this point of view it was legitimate, as mechanistic science attempted, to try to explain all properties of matter, life, and thought by the interaction of basic particles and the effects of the laws of physics and chemistry. It was common to hear it said that "more can never come of less" or that "time can bring nothing that has not already been given." From this came the theory of the epiphenomenal conscience and the impossibility of free will.

Scientific and philosophical thought have obviously evolved a great deal from these extreme positions. Today we admit readily to the increase in complexity that manifests itself in the course of evolution and to the emergence of new properties. However, we still have difficulty explaining the "vertical" transition from one level of organization to another level of higher complexity; from one "integron" (Jacob) to another integron; or from one "holon" (Koestler) to another holon. This does not mean that we shall never succeed, as the vitalists and spiritualists claim. Yet in spite of the sharp power of resolution of modern scientific thought, it seems difficult, because of the limitations mentioned, to interpret this vertical transition in any way other than by a juxtaposition of still positions—like the arrow of Zeno of Elea in flight or the arches of a bridge thrown across a river, which cannot follow the river's course (Bergson).

The Final Explanation: Convergence

The interpretation of the facts amassed by positive science can give a new meaning to evolution. Imagination, intention, and the poetic interpretation of reality help to reveal the full meaning of evolutionary facts. And personal motivation and the will to act depend in turn on the meaning that we give to events.

In this view each finality reaches ultimately toward a single end, located in the future, and in which it integrates itself. Every goal, every intention can be linked to a goal or an intention at a higher level and of a more general character. Finalities do not appear at the extremity where blind determinisms play. But at the human stage of evolution they are increasingly evident; humanity can handle its own destiny, thereby ensuring the relay of biological evolution. Everything points to the fact that evolution is converging toward a single end, which can be portrayed as a cone that is the reverse of the first. In this construction may be recognized Teilhard's cone of time, cosmogenesis, at the close of which the mind liberated from matter will be gathered at the end of time, at the Omega point (Fig. 87).

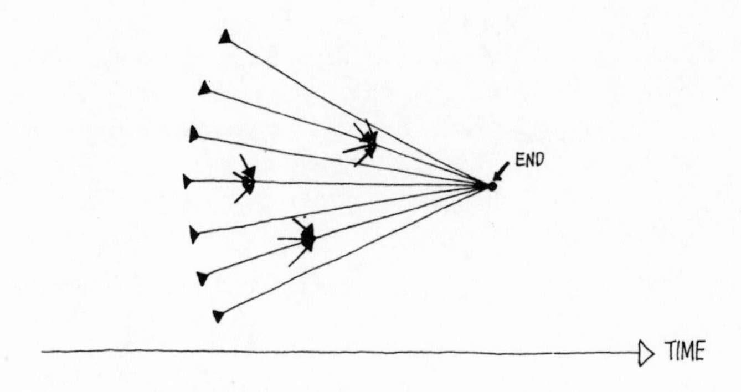

Fig. 87

Such a representation returns in reality to an inversion by the consciousness of the conventional direction of time. For the positive axis of cosmogenesis is defined here by increasing complexity (the increase of neguentropy). This new convention appears to be inseparable from finality as a method of explanation. But this is where interpretations will differ strongly: finality is not an *explanation* (this term ought to be reserved for the rational sense of exploration of the universe); rather, it is an *implication,* or an involvement. Final "explanation" belongs to an act of faith. It is no longer a "reverse cause" that forces evolution to execute a program established in advance or to follow the "design" of nature or of God.

In the finalist view generalized evolution appears as a movement that is antidispersive, selective, convergent, and creative of order—analogous, therefore, to any intelligent act. Contrary to thermodynamic evolution, which points toward conditions that are ever more foreign to us, convergent evolution would be directed toward what resembles us most; it would assume our values, our desires, our hopes. It would resemble an exploration and a conquest of an inner space-time confined "within" rather than an exploration and a conquest of an outer space-time dispersed "without."

This movement is by nature invisible to reason, which refuses to accept such a concept of evolution. It is not demonstrable; it can only be perceived, deduced, interpreted by the consciousness, which sweeps up, in the reverse direction, the facts amassed by observation and experience.

Are divergence and convergence complementary states? To represent both in a single diagram, one need only superimpose the two cones, for divergent evolution and convergent evolution are related to one and the same positive direction of time. This situation occurs even though the adherents of one approach or the other refer implicitly to what appear to be two contrary directions (Fig. 88).

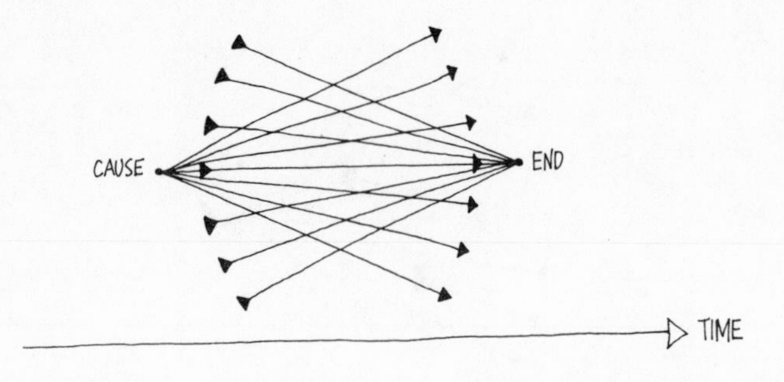

Fig. 88

Is there a dichotomy between these two states, or a complementarity?

In focusing on divergent evolution, one loses sight of its direction, its meaning, its finality. Human values, the subjective, the affective, the "meaning of life" have no place in the causal explanation—no more than do becoming, creation, or free will. The unquestionable advantage of the causal explanation is to be able to demonstrate its theories by scientific proof.

On the other hand, when one focuses exclusively on convergent evolution, the details of the underlying phenomena become vague. Even if one is utterly convinced of the direction or the meaning of evolution, of the interpretation given to facts, to events, or to the finality of every act, one has no proof to offer but "the evidence." And to want to demonstrate at any cost the "evidence" of convergent evolution is to miss the goal that one is trying to reach. In fact every demonstration points in the conventional direction accepted by the principle of sufficient reason. By using the approach and the language of science, one inevitably transforms into a divergent phenomenon what one believed to be convergent.

In superimposing the divergent cone on the convergent cone in the conventional direction of time, one rediscovers the status of complementarity of all phenomena that, subjectively or objectively, are linked to time. But in doing this, one projects the direction of creative evolution in the direction of one's individual future; there is an apparent reversal of time by the creative consciousness.

In observation, situations always precede representations (subjective models). In action, the representation of what one wants to do (the model of one's future action and its possible consequences) precedes the situations determined by this action (Fig. 89).

If the future of each life and the future of evolution coincide and are superimposed, it is because we imagine our individual future (and that of human society) as something "to be constructed," and therefore *before* action. We are in convergent time; its arrow points toward the increase of complexity. Perhaps confusion is born from the fact that we use the same time scale to measure the succession of events in our lives (from birth to death) and the stages in the life of humanity. The direction of historical time or evolutionary time should be the opposite of the direction of entropic time.

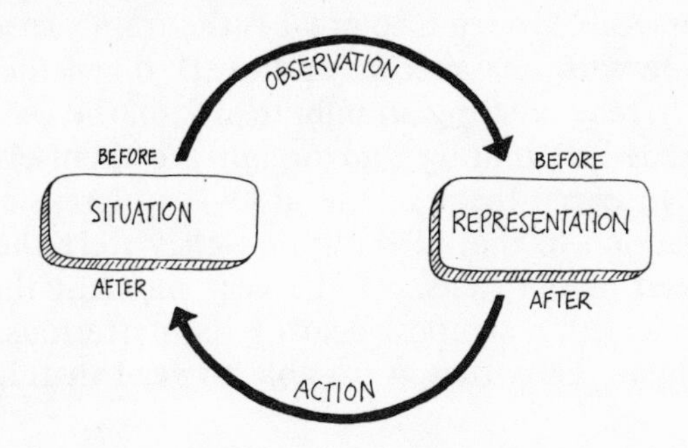

Fig. 89

Complementarity: A Third Route

To surmount these contradictions, we may use a third route, that of a complementarist dialectic inspired by the form of reasoning introduced by cybernetics. Like the systemic approach, this approach considers the totality of phenomena. Deliberately reintroducing the subject into the world of objects, it accepts a universe both perceived and lived in under two aspects, the subjective and the objective. Finally, it attempts to resolve the dualities and to go beyond the alternatives by laying down as its first principle the principle of the conservation of time.

To maintain the organization of an open system (living cell or human society) is to slow the speed of the increase of entropy in the system, or in this case to slow the passage of time. "Incapable of stopping the advance of material change, [life] nevertheless manages to delay it" (Bergson). *A fortiori,* to create information and organization, to compensate for the wear of machines, to use ways that make it possible to concentrate and channel energy, results in holding time, in preventing its being lost. It also contributes more effectively to slowing (and perhaps stopping in the intensity of the moment and not in the dilution of eternity?) the passage of time by balancing it against the creation of information. Time and information: two flows of equal speed, moving in opposite directions.

Conservation of time would then come about through the maintenance of a balance between speed of organization and speed of disorganization of the world. When evolution began, the flow of entropic degradation was preeminent. The activity of man, however, helped to oppose it with an increasingly intense flow of new information. We can illustrate this with a story. Living beings are passengers on an infinite train traveling at great speed, "the train of the second law of thermodynamics." Confined in small compartments, the passengers measure time by counting the signs that pass their windows with regularity. Intrigued by the inscriptions on the signs (impossible to read, so great is the train's speed), the passengers communicate with one another and break down the partitions that separate them, thereby creating an infinite aisle in the center of the train.

Having succeeded in uniting and organizing themselves in order to build machines to carry them in the aisle—at increasing speed but in the opposite direction to that of the train—they were then in a position to offset the speed of the signs. At the very moment that the speed of the signs was canceled, a signpost bearing its mysterious inscription appeared before them. Thus they were able to read in full the "secret of the universe."

Man resembles Janus, the god of two faces. He is the meeting place for two different qualitative perceptions of the direction of time. His life runs in the time of death, but his organizing action on physical and conceptual systems is in the time of the life of the world.

Through his actions each man transmits a part of himself into the universe. He fills a reservoir where something is being stored. Consciousnesses are (and probably will be even more effectively) interconnected and synchronized through means of communication in real time and by collective memory. This collective consciousness becomes informed by obtaining information on the universe (through research) and communicating it (through education). All creative action, at all levels of society, contributes in its own way to the organization of the world and its advancement toward higher levels of complexity.

The increase in complexity is neither unavoidable nor irreversible. All organization, no matter what its form, remains subject to degradation, to use, and to aging, whether it be living beings, machines, buildings, or information. Human society could even be destroyed instantly by nuclear catastrophe.

However, it is the individual creative action that compensates for the passage of time. For every original work is analogous to a reserve of time, to a *potential time.* Along with the concept of potential energy, then, we might propose that of potential time. The significance of the concept can be guessed: potential time is *information.*

Consider two examples, one at the biological level, the other at the level of society. The information necessary for the reproduction and maintenance of the structure of a living being is inscribed within the DNA molecule. This molecule represents all potential time amassed by the past evolution of life. The message is of high improbability; the actualization of this potential in the time of making copies will constitute the short span allotted to existence. The information that was present at the origin of this life will only be irreversibly degraded. Like the noise that covers and slowly blurs the meaning of a message, disorder sets in and increases. Entropy rises and errors accumulate. From reproduction to reproduction, from synthesis to synthesis, the organism ages, then dies. It has exhausted its "reserve of time," its reprieve has expired. It has attained its most probable state—death.

We see the opposite when we consider the life of humanity. The generation of information (potential time) in human society is accomplished at an accelerated rate as a result of the ceaseless efficiency gathered in storage and processing systems. As Gaston Berger has observed, humanity seems to grow younger.

Thus we can distinguish between the evolution of an individual life, which belongs to ontological time (the time of making copies), and the evolution of life that culminates today in the collective life of humanity, which belongs to phylogenetic time (the time of the creation of originals).

The dialectical approach proposed here accepts two complementary languages: that of reason, of scientific knowledge; and that of "meaning," of art, poetry, and religion. The scientific language (mathematics, physics) is rich in information and poor in human content, while the language of meaning (politics, religion) is poor in information but rich in human content.

Using the two languages, one can try to answer the "how" without neglecting the "why"—without separating the objective world from the subjective world. For they form the two complementary aspects of reality and knowledge, in spite of the enormous disproportions between the objective, physical universe and the subjective universe of individual consciousnesses lost in the immensity of space-time.

In the complementarist view, information and neguentropy are no longer divided into two separate worlds; they are the *hinge* between the objective and the subjective. Although they are superimposable and equivalent, information and neguentropy possess opposite "temporal poles." In fact neguentropy, the objective measure of information, is compelled to head (as soon as it is used) in the direction of entropic time. On the other hand, information, the subjective meaning of neguentropy, is compelled to head (once it has been acquired) in the direction of creative duration (Fig. 90).

Through observation and in the certainty of the tangible, we discover the world in a direction analogous to that of waves diverging from a source: the direction of conventional time. The universe now appears to us in its energetic, quantitative, material, and objective aspect. Through creative action and in the richness of living experience, we discover its

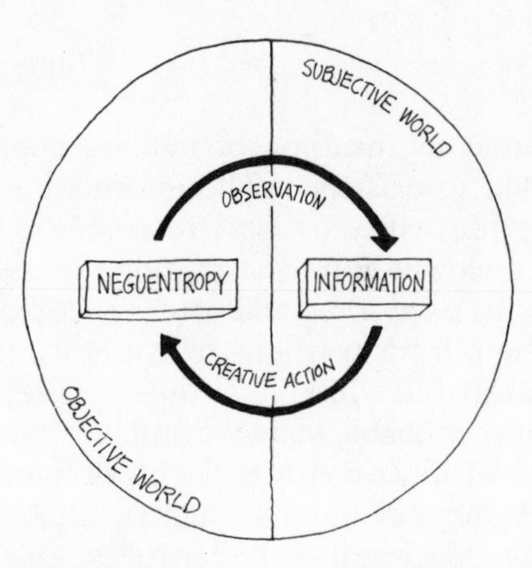

Fig. 90

other face in the direction of waves converging toward a center. It is the spiritual and subjective aspect through which the universe becomes more and more meaningful.

The two basic entities found at the end of this reflection, like the two sides of a single reality, are energy and the mind; their intermediate aspects are matter and form (or information). Yet everything appears as though only two things existed in the universe: *informed energy,* or matter, the fabric of knowledge; and the *materialized mind,* or information, the framework of creative action.

If there is conservation of time, freedom will be totally contained *in the present.* Thus, the universe appears as a consciousness that creates itself as it becomes conscious of itself. The trail it leaves, and which we observe, is the phenomenon of evolution.

3. EVOLUTION: GENESIS OF THE IMPROBABLE

Evolution is the history of self-organization of matter in increasingly complex systems. It is a very general process that includes prebiological, biological, and social evolutions, and for this reason the mechanisms most widely used to explain biological evolution (mutation and natural selection) are no longer sufficient. They must be expanded and generalized to make them applicable not only to biological systems but to physicochemical and social systems as well.

The global view of evolution, faithful to the systemic approach, integrates energy, information, and time. Its goal is to come to see in the same light the genesis of the organizations of life and society, their maintenance in time, and their evolution.

Darwin's explanation of biological evolution is based on three concepts: spontaneous variation, the struggle for existence, and natural selection.

Spontaneous variations are the random mutations produced in the chromosomes, which determine heredity. These variations generate new forms; thus there is an increase in the variety of forms present.

The struggle for existence results from the combination of two effects: the formidable reproductive capability of living organisms and the limitation of energy resources (or the dangers of the environment). The better-adapted organisms will survive and reproduce; those less well adapted will die. The outcome is simple in its severity: survival or disappearance.

Natural selection is the ultimate approval of the environment, which acts here as a filter. Reproduction permits the transmission from one generation to another of the ability to adapt to certain environmental conditions. There is reinforcement of the better-adapted species, and their populations increase. Each individual, being subject to mutation, has the potential to affect the entire course of evolution.

In order to extend this classic mechanism to the evolution of all complex systems, we must replace the three Darwinian concepts with *generation of variety, survival* (or *disappearance*), and *competitive exclusion.* Every evolutionary mechanism in fact rests on the combination of three elements: a random generator of variety, a system of stabilization (and therefore self-maintenance), and a selector.

I shall take up these three stages, genesis, survival, and exclusion, giving examples from physical, biological, and social systems.

The Genesis of Form

Thermodynamic equilibrium is death—monotony, homogeneity, the tepidity of entropy. Life, on the contrary, like all forms of organization or information, is a deviation from equilibrium, a temporary evasion, a reprieve.

The problem of the appearance of new forms (morphogenesis) can be illustrated by two questions: How can order, information, and variety be born out of disorder and homogeneity? How can one pass from a state of equilibrium to the "controlled disequilibrium" that is life?

The treatment of classic thermodynamic principles by the theory of information modifies radically our idea of equilibrium. A deviation from thermodynamic equilibrium is *the equivalent of information;* the expressions "far from equilibrium" and "recognizable in the environment" have precisely the same meaning.

Consider two examples, an iceberg and a sand castle. An iceberg floating in the sea is conspicuous in its environment. It represents structure, organization, and information. When it melts, each drop of its water mixes with that of the sea. Entropy is at a maximum and equilibrium is attained.

A sand castle is made of the same material as the beach. It, too, represents a deviation from equilibrium, and it has a form that is readily recognizable in the homogeneous environment of the beach. But exposed to the wind and the movement of people on the beach, it soon becomes lost in the environment; it disappears completely when each grain of its sand blends with that of the beach.

All organization is like the iceberg or the sand castle. The problem posed by morphogenesis is not far removed from that posed by the transformation of a small part of the beach into a sand castle. The two questions are now reduced to one: How does every deviation from equilibrium—every generator of form—make its start?

At the base of this deviation and its preservation in time is the effect of positive and negative feedback loops. Every deviation from equilibrium begins with a simple fluctuation, and this fluctuation can be amplified through the play of positive feedback. In order to maintain itself in time, the fluctuation must be stabilized by negative feedback loops, which give rise to prolonged oscillations and then to cycles. These are characteristic processes of the vital functions of self-maintenance.

That everything begins with a simple fluctuation is a fact that rests on a property well known to physicists: a system that is stable and homogeneous at the macroscopic level is no longer so at the microscopic level. Take the example of a crowd: seen from afar, it presents a homogeneous appearance; its overall behavior is predictable, yet the actions of individuals can create fluctuations around a state of statistical equilibrium. These fluctuations can broaden and lead to a new and unpredictable overall behavior.

The same is true for molecules, which makes their study especially interesting with respect to the creation of living forms and the origin of life. A population of molecules forms a stable and homogeneous system at the macroscopic level, but at the level of the individual molecules the system is no longer homogeneous. Collisions, reactions, combinations that make and unmake themselves represent the fluctuations out of equilibrium. *Each random fluctuation is a possibility for new organization.* It is a kind of information. Amplified by positive feedback, each fluctuation is a *random generator of variety,* found at the base of all evolution.*

A particular form of fluctuation that plays a fundamental role in the genesis of an organized structure is the autocatalytic reaction. There is autocatalysis when the products of a reaction serve as catalyst in the same reaction. An autocatalytic reaction can lead to the emergence of an ordered structure from a homogeneous system. This is the case in a chain reaction which produces (following random molecular collisions) a more complex molecule that is able to catalyze *certain steps in its own formation.* The chain closes on itself to form a positive feedback loop. From simple molecules present in the environment and acting as

* On positive feedback, see pp. 72 and 87.

building blocks, the complex molecule assembles itself. The process becomes faster and faster as products hardly formed accelerate the building process (Fig. 91).

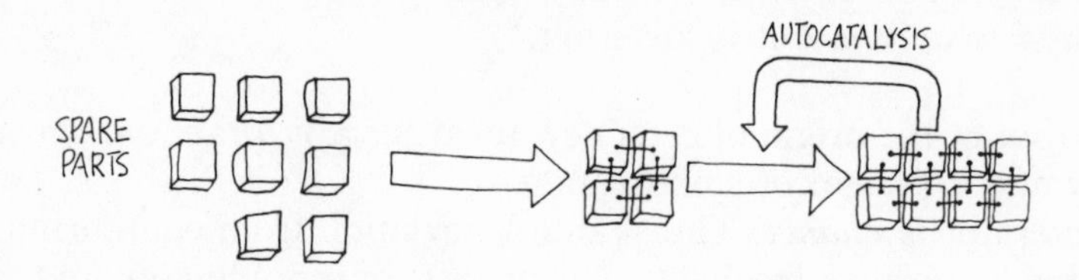

Fig. 91

At the molecular level this process is the equivalent of biological reproduction: to remake oneself faster than the original was made. The molecular species endowed with autocatalytic properties invades the environment. This explains the predominant role of certain molecules such as proteins and nucleic acids in the origin of living organisms.

Temporal asymmetry (see p. 166) is found at the level of molecular reproduction. A long period of time is required to produce the first catalytic molecule (the original), but once it exists it accelerates the steps that lead to the making of two, then four, then eight molecules of the same species. Thus copies are made quickly from spare parts present in the environment. This type of mechanism is also at the base of animal reproduction. Phylogeny requires a long time to produce a new species, but ontogeny permits the making of copies in a relatively short time. The demographic explosion—and its acceleration—is the direct consequence of efficient autocatalysis of the human species.

Fluctuations that prolong and amplify themselves can take the form of periodic oscillations in time. They occur, for example, when the presence of certain chemical substances causes the degradation of one catalyst and the regeneration of another—and vice versa. The concentrations of the two substances oscillate over long periods, moving to and from their minima and maxima. Similar oscillations are found in the relations between populations of predators and their prey. When *E. coli* bacteria and paramecia are cultivated in the same environment, the number of each colony oscillates between its maximum and minimum. The *E. coli* population is small in the presence of a large number of paramecia; then the latter die because they cannot find enough food—which enables the bacteria to reproduce rapidly again. And the process repeats itself.

Such oscillations represent the beginning of regulation through negative feedback, and they lead to stabilization. When autocatalytic chain reactions become extended in highly ramified networks, a branch may close on itself and form a *cycle.* Then the sequence of corresponding reactions becomes stabilized through negative feedback. Thus there is self-maintenance and self-selection. This explains why such cycles are found at the base of all life processes (cycles of cellular metabolism or ecological cycles).

Autocatalytic fluctuations, oscillations, and cycles can lead to the birth of *organized structures* out of disorder.*

At the root of the origin of each new form there is a random generator of variety and a system of stabilization.

The generator is *chance.* The slightest deviation from equilibrium can be amplified by positive feedback. The process of reproduction and mutation in living beings combines random generation of the most varied forms with their autocatalytic development. Environment, as we shall see, plays the role of the selector.

In social evolution unforeseen events, accidents, and environment-generated aggressions form the seeds of change. These events can be captured, selected, and triggered for political ends. Ideas, new directions that are the result of research and thought, are at the outset random fluctuations. They will be selected, saved, or abandoned depending on the play of rewards and reinforcements that link each person to the system that gives him life.

The system of stabilization and selection represents *necessity.* It causes the environment to intervene. This prevents the separation of an open system from its ecosystem. (Compare Figure 92 with the diagram on p. 66.)

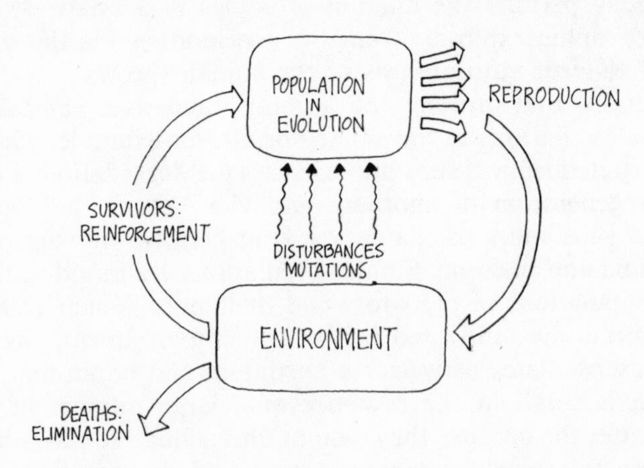

Fig. 92

* Von Foerster calls this process *order from noise.* According to I. Prigogine, it is order by fluctuation. Organized structures transforming energy are "dissipative structures."

The environment acts as a filter, keeping only the best-adapted forms. The penalty for not adapting is elimination and death. By upsetting homeostatic systems, the environment forces them to adapt and evolve. It is through the reinforcement loops (represented by the best-adapted survivors) that environment exercises its power of selection: obviously, only the survivors can transmit a favorable mutation to their descendents.

How does the selector operate? From a new organization, a new species, a new idea, how does evolution move? Toward growth, equilibrium, or decline?

Exclusion and Divergence

Autocatalysis inevitably involves rapid growth and acceleration—and conflicts with the environment. Growing systems drain energy for their own use; when the resources of the environment are limited, these systems enter into competition with others. Some survive, others are eliminated.

In this respect autocatalysis must be linked with self-selection. Natural selection must not be confused with an arbitrary "choice" effected from the "outside" by a supranatural entity or even by an environment endowed with some "design." The old concept of natural selection must give way to a more general concept that integrates duration and acceleration: *competitive exclusion.*

Competitive exclusion is based on speed of growth, acceleration through autocatalysis, and liberated power (see p. 104). Imagine two populations living in the same ecological niche and competing for limited resources. These two populations cannot coexist in perfect equilibrium unless their reproduction speeds are identical.* As soon as the rate of reproduction of one population exceeds that of the other, even by a tiny fraction, this population will have all the opportunities to eliminate its rival. The actual situation is clearly much more complex, for it involves the interdependencies of several populations.

Carried to its extreme, should not competitive exclusion end in a single species—the best adapted—selected at the expense of all the others? Human beings, for example? In fact such exclusive selection is impossible because it would destroy the ecosystem. Recall the law of requisite variety (see p. 87). The predominance of a single species or too drastic a reduction in the number of species present would cause a fatal disequilibrium. The ecosystem would not survive such a simplification, nor would, *a fortiori,* the systems that evolve within it. Self-conservation involves the

* It would be the same for a population of molecules displaying autocatalytic properties or a population of prebiological systems (the rudimentary ancestors of living organisms).

entire system—open systems in evolution *plus* the ecosystem.

Acceleration is one of the characteristic features of generalized evolution. Duration of time contracts from the first living forms to human societies. Human intellectual or sociotechnological evolution is even more accelerated than biological evolution. Every invention is the equivalent of a biological mutation. Man can invent and make a mistake without having to await the birth of a new generation to determine the results of his creations. In biology, to eliminate a useless invention one must always eliminate an individual. Moreover, the transmission of useful "biological inventions" is sequential; it happens only at the moment of transition from one generation to another. In intellectual evolution, however, what has just been invented can theoretically benefit everyone; the techniques of diffusion and storage considerably accelerate the sociotechnological evolution.

Linked to acceleration, competitive exclusion introduces temporal gaps that are difficult to fill, between two or more types of evolution. Now one understands the importance of the relative growth rate between two systems or two populations in competition in an environment of limited resources. Every increasing gap between two metabolic rates can also lead to the elimination of the slower one. This is as true for biology as it is for human society.

Thus the concept of *temporal divergence* seems to me to be basic to an understanding of the general mechanisms of evolution and "selection." Moreover, it has the advantage of linking closely evolution and time, which—paradoxically—scientists have been trying to separate (Fig. 93).

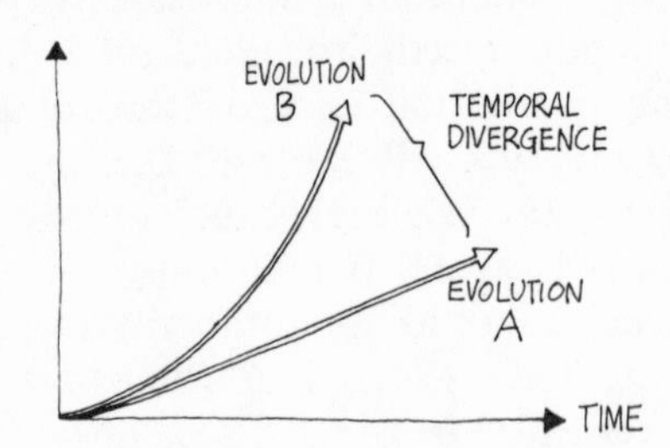

Fig. 93

So important a concept has not failed to have a profound influence on the development of the philosophical, economic, and political ideas on which modern societies are built. It is particularly enlightening in this respect to follow the direct line of ideas leading from Malthus to Darwin, from Darwin to Engels and Marx and the familiar concepts of "the struggle for existence" and "the class struggle." An economic law gave birth to a biological law that in turn was the basis for a new economic law.

In September 1838, some time after his return from the voyage of the *Beagle,* Darwin read Thomas Robert Malthus' *Essay on the Principle of Population,* which had been published in 1788. Suddenly he realized the fundamental importance of the temporal divergence between the rate of population growth and the rate of food production, which are the bases of Malthusian theory. "It at once struck me," Darwin wrote, "that under these circumstances favourable variations would tend to be preserved, and unfavourable ones to be destroyed. The result of this would be the formation of a new species. Here, then, I had at last got a theory by which to work."

The true driving force of evolution is the extraordinary power of reproduction of living beings. Potentially each species has the means for overrunning the world with its posterity; what prevents it from doing so is competition and death.

Engels was impressed with Darwin's ideas and realized the great generality of the evolutionary mechanism he proposed. On December 12, 1859, he wrote to Marx: "All things considered, this Darwin whom I am now reading is absolutely sensational . . . no one has ever made an attempt of such scope to demonstrate that there is an historic development in nature, at least no one has done so with so much success."

Marx, who lived in London, had the occasion to meet Darwin. In June 1862 he wrote to Engels: "What amuses me in Darwin, whom I have seen again, is that he *also* claims to apply Malthus' theory to plants and animals. . . . It is remarkable to see how Darwin recognizes in plants and animals his own English society, with its division of work, its competition, its opening of new markets, its inventions, and its Malthusian 'struggle for existence.' "

Among societies the law of competitive exclusion takes into account the widening of the gap between rich countries and poor countries. The unbridled consumption of energy in rich countries, related to the rapid pace of economic development, leads them to drain increasing energy flows from an environment that is becoming impoverished. In addition, growth and acceleration linked to the control of the regulatory mechanisms of a system of lower complexity lead to the domination of the weaker by the stronger. Beyond selfish interests there are moral, ethical, and humanitarian values that ought now to guide us. Without them we are in danger of seeing a phenomenon of inexorable competitive exclusion: the self-selection of the rich countries and the elimination of the poor countries. The catastrophic consequence of this will be the loss of an even more important treasure of humanity, the cultural and human variety necessary to its evolution.

Equilibrium and Zero Growth

For biologists growth is only a step toward equilibrium. Once attained, it is not static equilibrium but dynamic equilibrium. Static equilibrium, as we know, is death.

The concept of equilibrium in chemistry is based on Le Châtelier's principle: "If one varies the conditions imposed on a system originally in equilibrium, the equilibrium will move in a direction that tends to return the system to its original condition." This was a cybernetic principle before Norbert Wiener introduced cybernetics; it is regulation by negative feedback. This principle made possible the great laws that govern chemical reactions. But it relates to closed systems, whose evolutionary direction is determined by the increase in entropy. On the other hand, in an open system the direction of evolution is determined by the increase in information or—its equivalent—the decrease in entropy.* The stationary state that it maintains is comparable to a controlled disequilibrium, a flight forward. One is wrong to speak of "equilibrium of the inner milieu," "price equilibrium," "balance of payments," or "social equilibrium"; in an open system there are only controlled disequilibriums. This type of "equilibrium" is born from speed, like that of the surfer who leans forward to ride a wave that remakes itself endlessly beneath his board.

One of the best examples of "controlled disequilibrium" is furnished once again by biology. In the cell the manufacture of the cellular fuel ATP is taken care of by the chain of electron carriers (see p. 49). At the head of the chain are the energy-rich molecules extracted from food. These molecules have a strong "electron pressure"; thus they will tend to release their electrons. Each carrier is located at a lower "pressure" level than the preceding one, and the energy represented by the electrons runs from one level to the next like a waterfall. At the end of the chain the molecules have given away their electrons, their "pressure" has fallen, and they combine with water and oxygen. Over the entire chain, however, "equilibrium" is maintained.

Figure 94 illustrates this by means of a hydraulic analogy. In each tube (open at the top) water is maintained at a stationary level, provided that input flow is equal to output flow. The water levels are not the same, for the pressure differs from one tube to the next (it is weakest near the drain). The same principle applies to the "pressure" of the electrons in each carrier; it is weakest at the end of the chain.

The network of stationary states in the cell gives life one of its most

---

* If the creation of information exactly balances the production of entropy in a system, the system remains stationary—it does not evolve. If entropy increases, the system becomes disorganized and disappears. The direction of evolution (change) in an open system is therefore determined by the increase in information or organization.

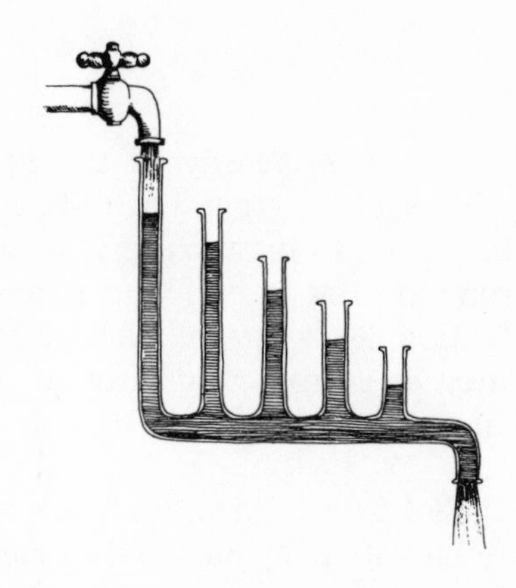

Fig. 94

remarkable properties: it maintains itself *at crosscurrent* to the flow of entropy. Incapable of overriding this flow, it balances it for a period of time.

It is in this respect that we must understand the expression "stationary economy" and the maintenance of a controlled disequilibrium. The expression is preferable to that of "zero growth," which introduces numerous misunderstandings regarding the finalities of economic growth. Zero growth is confused with a halt in the technological and intellectual progress of mankind or with a static equilibrium. The birth rate and industrial production are flows. The object of the stationary economy ought to be the maintenance of wealth at a desired level through the regulation of flows to their minimum output. To try to enlarge flows as though they themselves were wealth is absurd.

The Conquest of Time

Beyond the problems posed by pollution and the exhaustion of natural resources, economic growth "at any price" introduces a new constraint: it makes time a "consumable product." Time, like work, is broken up and rationed, for it is coming to be a commodity in short supply.

"The abundance of goods creates a shortage of time" (J.-P. Dupuy). We constantly lack the time to enjoy the objects we buy. Attached to each commodity is a "minimum duration for consumption": it takes time to read a book, to listen to a record, to watch television, to drive a car, to mow the lawn. "Time becomes a rare commodity in comparison with material things" (Dupuy). Its value increases with the standard of living—which accounts for the search for ways to take time away from chronophagic (from the Greek *chronos,* time, and *phagein,* eat) activities. People cut short the hours they allow for sleep, hygiene, meals, reflection, travel, family life, and sports. They prolong their working time in order to buy time-saving machines or to pay for other people's time.

Why save time? What deadline must we meet? Saving time without having a deadline (the *temporal dimension* of the goal to be achieved) leads, as we have seen, to wasted energy. The only way to "save time" on the human scale is to create, to organize the world.

Ways of fighting entropy are not to be found in accelerating the economic machine. Acceleration leads to an increase in consumption; it heads in the same direction as entropy and disorder. It would be better to look for ways to fight entropy effectively by *increasing the capacity for creation in society.*

Waste time to earn one's living or risk one's life to save time? That is the alternative that torments so many men and women in the industrialized countries. Instead of remaining trapped in this vicious circle, perhaps we could find in creative activity the ways for really saving time. To create something original requires time. The communication through education of this reserve of time furnishes everyone a "time capital" that one can use throughout one's life.

We must learn again to "waste time" in order to know better how to save it collectively. In our civilization of haste and waste the contemplation of a countryside, the conversation with a child, the participation in a sport, and even quiet meditation can seem a loss of time—but how many fruitful ideas, creative thoughts, and new hypotheses have been born in such moments?

The conflict between the unrestrained speed of our society and the awareness of a moment lived to its fullest has never been so acute. Our biological clock protests: the stress is too great. The organism reaches the limits of its resistance. Psychological time is no more taken into consideration than biological time, as many biologists and physiologists have pointed out. Time flows at different speeds for different ages in life. The young child, who uses a great deal of energy in proportion to his rate of metabolism, ages quickly. We could say that the younger the organism, the faster it ages. Psychologically that is expressed for the child in the slow passage of the outer time of reference, for he fills it with too many "priorities" of his biological time. The years pass slowly when one is young, then more rapidly as one becomes older. Everyone has this experience in the course of his life: the old man sees the child change constantly, but for the child old people remain the same.

In spite of this observation—which is confirmed by biological facts—the school day of a child is almost equal to the work day of an adult. The hours weigh more heavily for a child, but if he is properly motivated, he learns more quickly than an adult. Yet the child spends hours trapped in a world (school) that seems to him impoverished in comparison with the outside world.

Eager to save time, we live in an era known as much for its conquest of time as for its conquest of space. From communications and transportation systems to computers, we continue to invent machines for conquering time. But the machines may be a trap. Computers work to the nanosecond (which is to the second as the second is to thirty years); the quantity of information already present in the social organism is such that the lives of all persons living on earth today, laid end to end, would not make up a sufficiently long duration of time to receive, process, and assimilate all this information. The conclusion is simple: the amount of information needed for the functioning of society already greatly exceeds our capacity to handle it, even with the help of computers. That which is exchanged between one computer and another must be controlled; samples of these "conversations" have to be taken in order to allow decisions to be made. It is precisely this fraction, minute in comparison to everything being said in the world of bits and electronic impulses, that already saturates the processing capability of our mind.

What can be done? Shall we leave it up to the computers? Even in the Apollo program the entire hierarchy of computers checking in real time all the parameters of takeoff during the countdown was built in such a way that the final decision to launch was made by a human intelligence, that of the director of the program. The organization and the success of the Apollo program were due to the fact that it was a directed operation; those responsible were able to make choices, allocate time and resources, and organize time.

Our societies, however, still do not know how to choose their goals. To liberate time, to restore to everyone his free time, neither growth nor a stationary economy will be enough. We must succeed in setting clearly our goals and deadlines. Perhaps then we shall be able to fight effectively against a form of waste much worse than the waste of energy or raw materials—the waste of human energy. But in order to accomplish this shall we have to go so far as to overturn our scale of values? Goals and deadlines imply choices among many types of constraints. Every choice at the highest level is based necessarily on a hierarchy of values. Ours has lapsed; the failure of our industrial societies testifies to it. Can we discern in the new generation, more open to the global approach, the emergence of new values?

Six

Values and Education

Our education remains hopelessly analytical, centered on a few disciplines, like a puzzle whose pieces overlap rather than fit together. It is an education that prepares us neither for the global approach to complex problems nor for the interplay between them.

Nevertheless the present generation of eighteen- to twenty-five-year-olds itself poses problems globally. It seems that through a thousand parallel channels, passing from the traditional media to those of the counterculture by a sort of osmosis with nature and society, the young people have learned to discover for themselves a form of the systemic approach. In their own way they are applying it to the resolution of problems that previously defied the analysis and logic of their elders. Quite naturally they have taken advantage of the macroscope as a commando weapon.

Yet this emergent thought, this new manner of seeing and judging the world, is not the monopoly of one generation alone. Other men and women, of all ages and at all levels of society, share it today. Thus I prefer to call it simply the "new way of thinking."

1. BIRTH OF A GLOBAL VISION

The new vision of the world is not the effect of a single cause but the result of the convergence, integration, and interdependence of a large number of factors.

Some observers emphasize the catalytic effect of communications. Others underline the sudden realization, brought on by the environmental crisis, of the limited character of our planet. Others stress the clarity of the political critics of the industrial society and the analysis of its far-reaching ecological, economic, and human effects. And still others single out the postwar population explosion in the industrialized countries for having conferred a global "class consciousness" on an entire generation.

It is impossible to dissociate these elements from one another. However, we can try to distinguish in the global vision the influence of cultural and psychosociological factors as well as that of external factors associated with the phenomenon of evolution in its most general sense.

Certain great scientific discoveries have contributed, perhaps more than anything else, to expanding our vision of the world and to opening people's minds to the global approach. We must cite first the two most influential ideas bequeathed us by the nineteenth century: the idea of evolution in biology and the idea of entropy in thermodynamics. We have seen how they made it possible to integrate "vertically" the different levels of complexity in nature. But we must also cite new disciplines born out of the confusion of the 1940s: cybernetics, information theory, systems theory, and computer science. They insert themselves like wedges in the cracks between our partitioned but disjointed representations of the world, and they break apart our restricted and fragmentary vision of nature and society.

The plunge into the immense past of man, life, and the earth, as represented by the generalized study of evolution, leads to another confusion, this time of philosophical character. It traps such observers as us in our own objectivity.

Thanks to what science has taught him about the mechanics of its own evolution, the subject can now "put himself in nature's place," ask questions about the "logic of the living," and see himself in the "mirror of objectivity." The observer has cast his net over nature, thinking he could remain outside the phenomena he studies, a neutral, impartial, incorruptible spectator. But in his net he recognizes another aspect of himself, connected by his own fibers to the life and matter that preceded him on earth.

In this context we are led again to the eternal questions of our origin, the meaning of life, the consequences of our actions, our destiny. In the global context these questions appear in an entirely different light. The new vision no longer wants to—no longer can—separate object from subject. It cannot separate the certainty of the experiments accomplished patiently by science from the meaning and the finality of the conscious and creative action that transforms the world.

These upheavals in science and philosophy, joined to the major political ideologies of the nineteenth century (inspired as much by materialism as by spiritualism), have helped to facilitate the emergence of the new way of thinking. They brought to our lips the questions we are asking about the reasons, the motives, and the finality of our activity and our education.

Until very recently we were blind and deaf to the changes and the pulsations occurring in this great social organism of which we are the cells. We lacked the necessary perspective to discern its structure; we lacked the time to follow its slow transformations or to take apart its functioning machinery embedded in duration; we lacked instruments and methods for approaching the complexity of its organization and its processes.

Today, suddenly, everything is changed. The explosion in the means of communication, the acceleration of evolution, and the interplay of energy and economics have torn away the veil that hid the planetary totality from our eyes. At the same time the derisive and limited view of "spaceship Earth" made its dramatic appearance.

We see with our own eyes, with all the force of "live" broadcasting, the image of our planet as seen from the moon through the television cameras of the astronauts. A strange mirror: at the very moment that humanity regards itself, it could almost "wink an eye" by turning out all the lights of a large city.

This narcissistic vision prolongs and reinforces itself in time. Every day the newspapers publish photos, taken by meteorological satellites, of the cloud cover above the oceans and the continents. Geographical and geological satellites detect the smallest pollution of the seas—these lakes on whose borders we dwell—and send back pictures whose precision surpasses that of all the maps made from surface surveys.

Huge international organizations, travel agencies, airlines, hotel chains, and international expositions and sports events maintain a worldwide network that ensures the mobility of men and ideas. By sending us facets of our own image, these multiple mirrors force us to assume a global consciousness of both our diversity and our profound unity.

There is a close relationship between the speed of the diffusion of ideas and the collective perception of a "now" in the world. Fashions, the moral revolution, and technical breakthroughs spread with epidemic speed. Ideas have the "infectious power of viruses," as Jacques Monod has said. When the terrain is ready, it is invaded totally. The influence exerted by the young people is felt at once throughout the world—and on all the major problems: human rights, women's liberation, the protection of nature, economic growth, and the place of art, religion, and subjectivity in the industrial society.

The meeting of civilizations and cultures brings about an integration of the values of civilization and a complementary differentiation of cultural values. Through "world public opinion," "collective consciousness"—and even the "collective unconscious"—we see the outline of an emergent "psychology" of the noosphere gradually taking shape.

We never realize fully the importance of the vital functions of society until one of them slows down or we are deprived of them altogether.

Perhaps the best example of this simple observation is that offered by the energy crisis. Just as we discovered the pervasive role of energy in society, we realized suddenly the complexity of its distribution and utilization networks—for large industries, for small companies, for each one of us individually. As a consequence the worldwide interdependence of manufacturing industries and the interdependence of national economies were brought to light.

More important, we suddenly became aware of our power as individuals to act collectively through complex control systems over which we thought we had no influence. It was probably a revelation for many of us to discover the far-ranging effects on the economy of one country of the restrictions on automobile travel and the potential efficiency of systems for the salvage and recycling of discards—all dependent on the efforts of each individual.

The global perception of the functioning or malfunctioning of the social organism depends on many other positive and negative factors. Among them are breakdowns, such as the famous blackout of the east coast of the United States in 1965, and strikes of international impact, such as postal strikes, airline strikes, slowdowns by flight control operators at international airports, and walkouts by computer operators in banks. These strikes affect entire regions and countries, thereby reinforcing a sense of worldwide social interdependence.

Natural catastrophes such as droughts, floods, tornadoes, earthquakes, epidemics, and famines are as much emergencies as they are aggressive forces that make us consider, in spite of ourselves, the problems of others—and draw us more closely together, no matter what may be said.

Violence in all its forms—repression, guerrilla warfare, terrorist activities, airplane hijackings, the taking of hostages—mobilizes the attention of millions of persons simultaneously throughout the world. Nuclear testing in the atmosphere unites in opposition to it countries of very different customs and ideologies, and it focuses world opinion.

The economy, through its planetary dimension, contributes perhaps even more to this global perception of functions. The fluctuation of prices on the major stock exchanges, losses or lower prices in commodity markets, runs on gold, variations in the exchange rates among currencies, and the interdependence of problems that arise from food shortages, energy crises, and inflation—all these factors help to strengthen our sense of participation in the vital functions of an organism that surrounds us, that we do not see, but whose pulse of life we feel.

The acceleration of evolution culminates today in the still greater acceleration of the social system of the developed countries. The awareness of this acceleration contributes to the development among young people of a sense of impatience and a sense of the obviousness of the evidence.

Consider the sense of impatience. More than ever young people are measuring the gap between their expectations and the inertia of institutions, the disparity between what they learned in school, the world as they see it, and what the world could be. This impatience has been a part of every "new generation" whatever the period, but today it is heightened by the acceleration of events. Never has the future seemed so hazardous and so uncertain. Everything is possible: the collapse of the economic system, wars, dictatorships. Why then should one study patiently to accumulate knowledge that will be out of date when one is ready to use it? Why devote long preparation to careers that may be nonexistent in ten or twenty years? Why, as do so many adults today, live a boring existence in an office or a factory? Extreme attitudes often influence the responses to these questions: rather than try to integrate oneself into a social system that may be living out its last years, one should appropriate as soon as possible those things that one considers valuable, for nothing is guaranteed. We also see the beginning of a marginal class and marginal crafts, perhaps foreshadowing certain traits of the postindustrial society.

Consider, too, the sense of the obviousness of the evidence. For the acceleration of history is extraordinarily revealing. Like a film that has been speeded up, it discloses evidence of direction or purpose. Better informed through their many parallel channels of communication, observation, and mutual education—and more open, too—young people discern much better than their elders the occurrences, the developments, the situations that often escape the notice of experts and specialists. Each of these raw events is placed in a wider context that clarifies it and gives it its true meaning. The sense of priorities intensifies the feeling of impatience in the face of inaction, impotence, or resignation.

Could it be such a connection between a more global vision of the social organism and a more acute perception of the effects of acceleration that confers on the new way of thinking a gift of insight? In place of the ability to analyze, there seems to have been substituted a new faculty, that of pattern recognition. The new way of thinking appears to have taught people to view the world through the macroscope, to detect and recognize its grand patterns. This global vision has led to the substitution of the systemic for the analytical, shared subjectivity for noncommunicable objectivity.

The extensive modifications of our representations of the world are produced with such rapidity and lead to collective movements of such breadth that we confuse them most of the time with fads. Most social scientists neglect to study them or to analyze them by traditional methods. Only a few sociologists and some enlightened journalists have known how to integrate, by means of a more global approach to problems, the new dimensions of a world in acceleration. This has helped them to explain these changes better and to place them in the context of the general evolution of customs, values, and culture.*

The contribution of these sociologists or journalists has very often been criticized by their more traditionalist colleagues. For their approach is very different; like the generation they observe, they use the same faculties of *pattern recognition.* They look at things through the macroscope, using a systemic approach that allows them to integrate facts in multiple facets that enlighten their words and reveal their meaning. Some of their interpretations inevitably contain weaknesses, gaps, even errors, but their contribution complements that of the classical sociologists.

Today the "marginal" sociologists exercise a strong influence on a generation that recognizes itself in the mirror set before it. Their vision catalyzes and strengthens a universal movement whose breadth surpasses current fads and poses clearly the real problems of civilization.

2. THE EMERGENCE OF NEW VALUES

One very profound criticism of society and the nature of human relationships is elaborated in the turmoil of modern society. This is chiefly the fact of a generation that is often as foreign to traditional customs and values as the inhabitants of another world would be, were they suddenly dropped in our midst. It is difficult to regroup the principal criticisms and to identify the basic values on which the new way of thinking rests. Nevertheless I should like to try to do this, but not without taking some precautions.

These new values are not destined to be substituted abruptly for the old; there is no linear or sequential evolution here. But there is juxtaposition, coexistence, and sometimes complementarity, according to the degree or the speed of evolution of the various social groups for which the new values are relevant.

In attempting to answer the question so often asked by the generation in power—"What do the young people of today offer in place of what they are trying to destroy?"—I shall consider the major criticisms that the new way of thinking directs at contemporary society. Later in the chapter I shall bring them together in a summary table that stresses the main points of transition between traditional values and emergent

---

* I am thinking particularly of the work of Jacques Ellul, Marshall McLuhan, Herbert Marcuse, Margaret Mead, Edgar Morin, Charles A. Reich, Jean-François Revel, and Alvin Toffler (see bibliography).

values (see p. 205). The character of this presentation may be somewhat schematic, but its purpose is to allow a reexamination of the relationship between the aspirations of a generation and the education we propose for it.

Criticism of Authority

Criticism of authority is linked to criticism of the legitimacy of power as we perceive it. This power is symbolized by the "nine pillars" that, since the beginnings of civilization, have maintained law, social and moral order, and security in human societies: the State, the Church, the Family, the School, the Courts, the Military, the Police, and—more recently—Business and Medicine. Authority manifests itself in the orders (or resolutions invested with the authority of whoever holds the knowledge or the divine right) of "the chief": the president, the clergyman, the father, the teacher, the judge, the general, the policeman, the boss, the family doctor (Fig. 95).

Fig. 95

These guardians of the moral and social order were accepted by earlier generations in the same way that the institutions they represented were accepted. None was disputed; everyone accepted their commands and deferred to their authority. Today, however, we question the legitimacy of power and its applications. To action through influence and motivation, we oppose the direct exercise of power. For influence implies freedom of choice rather than physical constraint. And there are roughly just two ways of moving things and people: direct exercise of power and influence.

The direct exercise of power can depend on physical force or on psychological and moral constraints. Its legitimacy rests in the power conferred by the possession or control of an energy or money capital. Influence depends on the force of idea and example; it often has recourse to "knowledge capital." The first approach generally acts on structures in order to change people, and the time between the application and the result is relatively short. The second approach tries to modify people's minds in order to change structures, and the time involved is obviously much longer.

Great leaders are perhaps those who know how to measure out the two approaches in the light of the time and the circumstances. Totalitarian dictatorship would thus seem to be the political caricature of the direct exercise of power, while the caricature of indirect action through influence might be the form of intellectual dictatorship practiced in some elite educational and religious orders.

The new way of thinking questions all forms of the abuse of power. Attempting to avoid abuses, the new way of thinking tries to oppose institutional hierarchy and the centralization of power with the continuous evaluation of a hierarchy based on competence and the decentralization of responsibilities. The traditional pyramid of authority, rank, discipline, and domination is transformed into a more "horizontal" organization that resembles a living cell. In this type of organization, power, elitism, sense of duty, and adversary relationships are replaced by shared obligations, participation, interior motivation, and partnership. This is the reverse of traditional power and authority; it is management by the base, by interdependent communities. This reversal is foreshadowed in the proliferation of words prefixed with "self-" or "co-" whose power lies in the ability to evoke action: self-determination, self-management, self-discipline; co-ownership, co-responsibility, cooperation, and co-decision.

Criticism of Work

The essential criticism of work goes beyond simple disrespect for its value, and questions its ethic. It attacks the conditions, the environment, and "the rules" of work—not in order to praise idleness (as one might often think, judging from some extreme attitudes) but in order to liberate working time so that each individual may again govern his own time, work under his own conditions, at his own rhythm, and for irregular periods—so that each person may personalize his work.

Why produce so much when we no longer have the time to consume what we have produced and when what we produce irreversibly degrades nature? Why work so hard to accumulate material goods if we no longer have the time to fulfill ourselves in our relationships with others?

This criticism has repercussions for a whole set of conformities, practices, and rules that have always been taken for granted: diplomas, the career, competition, success. And it brings out the hypocrisy of the "work alibi."

Diplomas. No longer considered the keys to social success, diplomas are the means of defining, in terms of one's own deadlines and one's own potential, the personal rules by which one learns to organize oneself, to enrich one's mind, and to see oneself through. After the confrontations of the late 1960s, the students of the 1970s appeared to be concentrating more than ever on their studies and their diplomas. They became *self-centered,* they worked *for themselves.*

The career. It seems delusive to spend a large part of one's life preparing for duties that will no longer be the same when one is ready to assume them. Instead of a single linear career, one may prefer a series of multiple trajectories. One may even interrupt one's professional preparation for a period of reflection and commitment.

Life is a succession of *choices* and *objectives.* Adaptation is the rule. In a changing environment the laws of cybernetics bring out the efficiency of servomechanisms, which are capable of adapting themselves, and the failure of programmed mechanisms. Because of the acceleration of evolution, no career can be programmed. The choice of a career should no longer be the major decision of one's life.

Competition. Up to now professional competition has appeared to be a healthy motivation for success. The new way of thinking rejects all competition that is heir to the traditional "struggle for existence" and spurns any notion of simplistic comparison founded on "excellence" and "merit." For such comparisons generally lead to the arbitrary classification of individuals and to value judgments that limit and impoverish human relations.

It is the refusal to join the rat race. To finish or simply to hold one's own in this race requires the elimination of every human obstacle that appears before one. Today many people are rejecting competition for the "marginal" professions, where one finds now and then the warmth of human relationships and the time for reflection.

In a somewhat naive manner, society, freed from the notion of competition, no longer sees itself as a jungle but as a community of interests whose evolution depends on helping one another, on cooperation, shared education, and *partnership.*

Success. Social success, too, has long been considered the principal motivation of the professional life and an indirect factor in economic and social progress. With success come honors, attention, respect, position, security, material well-being, and power. These are the essentially selfish values of a civilization founded on the conquest and domination of nature and the servitude of one man to another.

In the new way of thinking, success is based on *personal accomplishment.* It is the enrichment of experience that one feels in one's contacts and interactions with other persons and other cultures, the pleasure that comes from work well done, the sense—still so difficult to achieve in the context of our societies today—of the usefulness and the effectiveness of one's actions. People are looking for a "role," an involvement, a cause, rather than the specialized but ultimately insignificant job that modern society all too often offers.

The hypocrisy of the "work alibi." Perhaps even more than it criticizes the diploma, the career, competition, or success, the new way of thinking criticizes the hypocrisy of rules stemming from work that has become an end in itself, that produces its own immaterial rules and develops a logic unrelated to real life.*

Outward signs of wealth are deservedly taxed, but we exaggerate the value of "outward signs of work." In many organizations, especially in Europe, one is still judged—and promoted—on the basis of the thickness of one's reports, the quantity of notes, memos, and letters produced daily, the number of meetings and telephone conversations, and the length of one's working day. Each of us in his professional life has known at least one incompetent boss who was incapable of making a decision or motivating those who worked with him but who nevertheless held on to his position because, raised in the same environment, he had the same values and the same ethic as those who admired his "work pattern" and his "devotion to duty."

We confuse feverishness and efficiency. Our remaining methods of control are based more on the quantitative than on the qualitative, which is harder to evaluate.

Already the image of the tired businessman or the overworked executive is no longer a source of respect but one of pity. Their excessive activity and the pressures they must endure are in many cases justified, made necessary by the responsibilities or the special conditions of a particular situation. But doesn't this activity often mask marital or family problems that one is trying to forget or to escape? "Exaggerating the value of work can hide a flight from reality," Denis Vasse points out in *The Time of Desire.* "Work can be the most deceitful alibi of man"; "the need to work lends itself to any subconscious justification." And there are the usual clichés: "I can't take a vacation, I'm swamped with work"; "I don't see my children because I never get home before nine"; "He's a slave driver, but you have to excuse him—he is overworked."

---

* Obviously this criticism and rudimentary analysis applies chiefly to salaried work symbolized by "the office."

These are no longer acceptable excuses, for they denote a refusal to accept full human responsibility. And many young people today are declining to squander their energy in a sterile and empty contest in which appearance has the advantage over reality, where the image that one creates counts more than what one actually does.

Criticism of Reason

It was in the name of reason and logic that political and industrial leaders influenced by scientific and technological achievements created the civilization of progress, economic growth, and the domination of nature.

The new way of thinking distrusts reason and logic. Of course the analytical method, Cartesian logic, and the principle of sufficient reason have been indispensable tools in man's attainment of a certain level of development; everyone recognizes this. But these methods, principles, and postulates are no longer the only bases of knowledge. To objective knowledge we can now oppose subjective experience; to "life" defined in scientific terms, the experience of having lived and the quality of that experience.

To emphasize the necessity of such an advance, I offer as example the replacement of the logic of exclusion (to which our education has accustomed us) by the logic of association. The logic of exclusion leads to reasoning in opposing and mutually exclusive terms such as true or false, good or bad, black or white. It leads to the well-known dichotomies of thought in which certain ideologies inherited from the nineteenth century take refuge. The class struggle (in the Marxist view) and economic competition (in the capitalist view), for example, opposed as they are, are in fact two sides of the same coin. Both are derived from the Darwinian concept of the struggle for existence, and this struggle is everything or nothing, life or death. On concepts of this kind we build a scale of values that determines our action or our opinion with respect to others: if I am right, you are wrong; if I win, you lose. This is the *zero sum* in game theory. It leads, as we discover every day, to sectarian and intransigent attitudes.

Biology and ecology show us that there are no such entrenched oppositions in nature. Every relationship or equilibrium is founded on pluralism, diversity, mutual causality. There is no logic of exclusion or opposition, but there is a logic of association or complementarity. Thus biological or ecological thinking leads to the emergence of associative values that foster tolerance, the respect for other ideas and cultures.

The logic of exclusion, associated with a causal, analytical, sometimes reductionist conception of society and its evolution, has led numerous political and industrial leaders to be concerned only with objects (things, people) and to disregard subjects (persons, life). We train men to manufacture material objects or to manage people conceived as objects situated in large organizations.

Today this basic criticism also applies, in a vague way, to technical progress, to the finalities of research or economic growth. From this comes the extreme antiscientific, antitechnological, antirational attitude that one finds on so many university campuses.

We look for a social role for science. Great universities traditionally dedicated to teaching and research now add "services" to their agenda, helping—for example—municipalities or government agencies to approach social problems of such complexity that their solution warrants multidisciplinary cooperation.

Such criticism of reason often leads to attitudes that are now and then extreme or naive. Yet these attitudes indicate a willingness to be open toward subjectivity—what observers sometimes interpret as an "escape" into the irrational or into mysticism: the infatuation with Oriental religions, astrology and magic, the rediscovery of Jesus, and even a sort of ecological pantheism verging on an "ecocult," or a devotion to the great cycles of nature.

Criticism of Human Relationships

It seems intolerable to the new way of thinking that those who trust in an all-powerful authority, in the value of work, in reason and logic, can, in the name of law and order, cover up crimes against a country and allow repression, hypocrisy, lies, and the manipulation of consciences to be used widely. The most recent history has shown us how the very people who deplored the erosion of traditional values, the lack of idealism in young people, the disintegration of manners and morality, were the first to corrupt and pervert human relationships in major government bodies or in business, through their untruths and their egotistical, partisan attitude.

Seen in such a context, the pressures exerted by American youth on the media and on Congress in order that justice be rendered in the Watergate affair could be the transposition to the level of political morality of the struggle they led against big business at the end of the 1960s to establish a new environmental morality.*

Human relationships, at all levels of society, ought to be founded

* Traditional morality (in politics, for example) was founded all too often on such principles as "the law of the stronger," "the end justifies the means," "out of sight, out of mind," and "never admit anything."

not only on a morality for individuals but on a new morality for the groups among them, one compatible with that for individuals. This group ethic, an essential intermediary between the morality of the species and that of individuals, has yet to be created.

One way of creating the new ethic is to go beyond one's own interests in order to understand other people better. The only real way of communicating with others, according to Charles Reich, is to be true to oneself first. One must succeed in defining one's own values, goals, and life-style, at the same time accepting and respecting those of other people. This point of departure brings out dramatically the poverty of human relations in contemporary society—the injustice, the privilege, the intellectual and material conformity, the segregation of old people, the lack of affinity between education and real life.

Criticism of the Plan for Society

Clearly we must try to go beyond the forum of the traditional political critics of society to indicate the directions that the new way of thinking is taking. Once again extreme attitudes and caricatures run the risk of concealing meaningful movements.

It has become almost commonplace to proclaim the failure of two plans for society, that offered by unrestrained capitalism and that offered by bureaucratic communism. But it is more difficult to define the "third direction," in which the new way of thinking is headed. Beyond the Chinese model, beyond Illich's conviviality, beyond ecologism, or Naderism, how can we integrate into the postindustrial society what each of these directions brings, while avoiding being taken in by the ideology under attack?

I am aware of the difficulties of the task, and following this chapter I shall try to describe in broad terms the likely structures and functions of such a society, using an indirect method, that of the "scenario." But at the critical level we must first reconcile the fundamental points that the new way of thinking will emphasize: centralization of power, bureaucracy, the descending flow of information, growth and consumption, the quality of human relations, dogmatism in science, anarchy in technology, the inadaptability of institutions, and deficiencies in education.

It is an account of failure—the failure of the application of science and technology to make us masters of nature, the "failure of the dreams of Descartes and Faust," as Roger Garaudy has said so aptly.

To formulate a new plan for society, we must start with new relationships among men, between man and nature, and between man and his future. We must call on the creative talent of the individual and respect of his independence, his pursuit of happiness, his search for pleasure, and his desire for personal accomplishment. Inevitably this requires, alongside the traditional "liberty, equality, and fraternity," pluralism, personalization, responsibility, and participation.

To clarify the values on which a new plan for society might be founded, I offer the following table, which contrasts traditional values with emergent values. Of course this is not a matter of mutual exclusion, but one of complementary enlightenment.

Traditional Attitude	*Emergent Attitude*
CRITICISM OF	AUTHORITY
Authority founded on power, secrecy.	Authority founded on influence, openness of motives, competence.
Respect for institutional hierarchy, devotion to established institutions, sense of duty, sense of obligation.	Continuous evaluation of a hierarchy based on competence, institutional innovation, personal motivation.
Elitism and dogmatism, centralization of power, conflicts of powers.	Participation, openness, criticism; decentralization of responsibility, relations based on competence.
CRITICISM	OF WORK
Importance of diplomas; responsibility based on age, theoretical knowledge, social rank.	Importance of participative experience; responsibility founded on ability to resolve problems and motivate people.
Linear career, programmed progression, competition, honors, success.	Multiple careers, succession of choices and objectives; cooperation, personal joy, personal accomplishment.
Prizing of contribution and personal effort, hard work, devotion to an organization; exaggerated valuation of "outward signs of work."	Prizing of creativity and collective merit; creative work at one's own rhythm, commitment to a cause, efficiency in accomplishing a given objective.
Material job security, need for hierarchical domination and discipline; specialized jobs.	Liberty achieved through acceptance of risk and through diversity of functions; need for cooperation and communication; "role" of social responsibility.

CRITICISM OF REASON

Logic of exclusion (Manichaeism); unidirectional, causal, sequential.	Logic of association (ecosystemic); mutualist, globalist.
Principle of sufficient reason, postulate of objectivity, analytical method.	Contribution of shared subjectivity, complementarity of objective facts and lived experiences, systemic method.
Pure knowledge.	Inventive thought.
Defense of finalities in science and technology.	Criticism of finalities in science and technology.
Acceptance of technical progress, economic growth and power, the domination of nature.	Acceptance of technical progress as a function of social need, equilibrium and sharing, partnership with nature.

CRITICISM OF HUMAN RELATIONSHIPS AND THE PLAN FOR SOCIETY

Partisanship, stubbornness.	Tolerance.
Aggression, cynicism, skepticism.	Openness, naivete, enthusiasm, sense of usefulness.
Use of others for personal ends; projecting an image of force and strength.	Respect for others; being true to oneself.
Domination, private interests.	Cooperation, sharing of interests, search for a group ethic.
Uniformity, homogeneity.	Pluralism.
Quantitative.	Qualitative.
National power, individual well-being, economic growth.	National involvement, individual improvement, balance and sharing.
Patriotism, chauvinism, nationalism, imperialism.	Internationalism, interdependence of nations and cultures, religious and philosophical contributions.
Unrestrained capitalism, bureaucratic communism.	Conviviality, leftism, Maoism, ecologism, radicalism.

A collection of new values does not make a political ideology. The emergent values could very well be rearranged; for example, at the individual level (morality, ethics, religion), at the cultural level (philosophy, science, technology, art), or at the political, economic, and social levels.

Such a regrouping would place them in a hierarchy and bring out their dominant values, but it could also lead to repetition and would require too normative an approach.

We must not founder in a smug idealism, seeing the remedies for all our ills in the new generation. The important thing is to notice—in the absence of any clearly recognizable manifesto, politics, or practice—how its ideas and its values modify modern society. More than through a coherent and even shared political approach, it is through example, influence, individual action, collective movements, life-styles, and behavior that the new way of thinking will slowly but profoundly change our industrial society.

On the other hand, what appear as independent actions or fashions belong in the coherent global context that I have tried to describe and explain.

We are also witnessing the birth of a new religion founded not on "revealed truth" but on a truth compatible with our objective knowledge of the world. It is an emergent religion that results from a collective creation and that accepts the immersion of the spirit in matter.

In all their diversity, the ecological movement, the search for a spirituality drawn from Oriental religions, the T'ai Chi movement, and the human potential and awareness movements that still flourish on the California coast all denote a search for a global vision of the universe that is compatible with a personal ethic and individual and collective action.

We find the most exaggerated contrasts at the level of life-styles and behavior. To make them more personal, we might enjoy comparing (admittedly a somewhat simplistic contrast) the "technocratic" style and the "hippie" style of the 1960s. What could be more revealing than the study of these two extremes? For technocrats, the only thing that matters is action—doing and making do, through reason and technology. For hippies, what counts is feeling, relationships with others. One can contrast the two attitudes by referring to the discussion of the various criticisms and the table of the preceding pages. One will find, in varying degrees and with many shades of meaning, this characteristic contrast of the two life-styles.

Are we moving toward a schizophrenic world divided between those who believe themselves invested with a mission to push the world forward and those who prefer to profit from it—between robot actors and pleasure-loving spectators? To avoid such a division, we must be more attentive than ever to the gropings of the new way of thinking. We must define, along with the young people of today, free of all demagogy and paternalism, the main directions of the education that they need to face the twenty-first century.

3. SYSTEMIC EDUCATION

The emergence of new values changes each personality even as that personality is modifying its relations with others and with the world. What does modern education propose in the face of the demands of an entire generation? Far from helping it adapt to a new environment, will it constitute a cultural aggression? Or, as Marshall McLuhan calls it, a new "tribalization"?

For some years efforts have been made through traditional instruction to integrate the disciplines and to increase motivation and participation among students. Yet new methods and techniques still do not offer a global approach based on the systemic approach. That is why I want to try to outline the basic principles of a systemic education and to suggest several new approaches that might be integrated with traditional instruction.*

One way to evaluate what a systemic education can achieve in comparison with traditional instruction is to start with an extreme, almost exaggerated point of view, that traditional instruction is based in part on principles and methods inspired by those used to increase productivity in shops and factories. In education the division of work is replaced by the division of knowledge. Thus we can appreciate the limits of traditional instruction, its approach, its means, and its methods. This instruction emphasizes, essentially, seven principles that I might define in a very irreverent manner:

Basics: Knowledge that one must master before knowing how it will be useful.

Subjects: That which each of us must assimilate in small quantities in order to acquire a "minimum knowledge."

Program: Organization of subjects in time in order to increase the efficiency of the process of acquiring knowledge. (That which is not in the program obviously has no educational value.)

Course length: The theoretical minimum time needed to assimilate a given quantity of information.

Equality: The principle that says that everyone shall receive the same amount of information in a given time (too bad for the slow ones, too bad for the bright ones).

Fields: Processes of "fractional distillation" in which each plateau represents a school year and through which an individual has to specialize for his entire life.

Examination: An initiation rite invented by adults, in the course of which the student demonstrates (so that he may quickly forget) what he has temporarily learned, in order to obtain in exchange a passport for entry into active life, called a *diploma.*

In all countries, clearly, efforts are being made to bring greater flexibility to this rigid framework of instruction. Numerous innovations have

* The principles of this approach pertain to various levels of instruction: primary, secondary, advanced, continuing, and adult.

already upset traditional methods. But without the global approach the various attempts to modernize instruction are perhaps doomed to failure. Among the most striking innovations are the audiovisual method, the multidisciplinary approach, teaching machines, programmed textbooks, and computer-assisted instruction. Each of these means is often considered an educational innovation *in itself,* suitable for accelerating the process of acquiring knowledge, and *therefore* an efficient means. Yet if one were to evaluate the systemic impact (in relation to all other forms of instruction) of each of these methods, one would see that their incorporation in an educational process that remains basically unchanged does not lead to the "educational revolution" that people expect.

The Illusions of Educational Technology

The audiovisual method has immediate pedagogical utility only to the degree that the student himself repeats what he has just seen on the screen. (Piaget noticed this long ago.) That is how the student registers new facts. Knowledge is not a "carbon copy of reality"; it is an "operative process" that ends in transforming what is real into action or thought, in acting on objects in order to transform them. Thus the feedback loop between observation and action, of which I have spoken repeatedly, ought to be found again at the instructional stage. Without feedback, isolated audiovisual instruction risks being only "another verbalization of a picture." To avoid this, the audiovisual presentation must be filled out with individual action, group participation, and the simulation of reality.

The purpose of the multidisciplinary approach is theoretically to permit the solution of complex problems by benefiting from the illumination of several disciplines and the complementarity of their methods and techniques. But without a systemic approach to blend and integrate the respective contributions of each discipline, the multidisciplinary approach never goes beyond the mere juxtaposition of disciplines. A true multidisciplinarity cannot arise from the *a priori* juxtaposition of specific disciplines on the same campus or in the same university building; it must be the result of a purposive organization, made necessary by problem solving. Experience shows that multidisciplinary cooperation is more effective in systems design (convergence of disciplines) than in systems analysis (divergence of disciplines). When we disregard these facts, we only mingle the researchers of various disciplines, believing we are leading them to collaborate. We create a disparate organization, not a judiciously integrated functional system.

About fifteen years ago educators (and private companies) had great hopes for "educational technology": programmed textbooks, teaching machines, and computer-assisted instruction. Thanks to the works of

Skinner and Crowder on programmed instruction, the techniques of linear and branched programs allowed professors to dissect their courses and to give their students a "predigested" instruction that was entirely new.

But the use of the new techniques was never as important as their promoters had predicted. Programmed texts in general had little attraction—as little for those who edited them as for those who used them. As for the teaching machine, it was too costly to be used on a wide scale, and it did not adapt well to student work habits.*

The computer represents, in principle, the ideal extension of the programmed text and the teaching machine. We can write programs, simulating a dialogue between teachers and students, that broaden the scope of programmed courses and give them more flexibility. Since 1960 experiments in computer-assisted instruction have been made in several American and European universities. They have shown that it is possible to use the computer to individualize instruction, to control a multimedia environment, to communicate with hundreds of students at the same time, to test and mark students, and to suggest further reading to supplement their knowledge.

Unfortunately, the cost of computer-assisted instruction at present is much too high; systems in operation are too expensive to be merely "turning the pages of a good programmed book." And the results that have been achieved in student interest and educational effectiveness have been little more than modest. Today, apart from some significant results, we cannot say that computer-assisted instruction represents, for the moment, a direction as promising as we thought it at the end of the 1960s.

The relative failure of these new educational methods emphasizes the need to decentralize technical means and to increase student participation. It is clear that the new educational technology follows in the direct line of the traditional unidirectional courses, of which it is more often than not only a straight technological transposition.

The Basis of Systemic Education

The systemic approach in education cannot be substituted for the traditional approach, nor can it resolve magically its principal problems. The systemic approach is an indispensable complement to traditional education. But the effective implementation of this complementarity necessitates both a simplification and an enrichment of present-day instruction. A simplification of our instruction because if we continue in the analytical approach, there will be (there is already) *too much to learn.* And an

* Programmed texts and teaching machines are nevertheless used regularly today, and some editors have succeeded in preparing excellent manuals for specific applications.

enrichment because the systemic approach, uniting facts in a coherent set, creates a conceptual frame of reference that can facilitate learning by traditional methods.

Systemic education must also define its principles and its methods by beginning with biological facts and psychosociological fundamentals—not to impose a certain kind of education that would be the same for all students, but on the contrary, to help all people, whatever their age or educational attainment, to acquire new knowledge and to make more effective use of it. It seems to me that systemic education should try to benefit, more than it does at present, from our knowledge of the functional organization of the brain and the basic components of human nature.

Recent research on the organization of the brain has revealed a pronounced functional difference between the two cerebral hemispheres. Because of the reversal of the nerve fibers that occurs in the *corpus callosum,* it is the left side of the brain that controls the right side of the body, and vice versa. And it is the left hemisphere of the brain that controls verbal activities such as reading and speaking, while the right hemisphere controls the perception of spatial relationships and pattern recognition.

The solution of problems requires two kinds of cerebral functions. The analytical function processes information sequentially; the intuitive function processes information simultaneously.

In other words, the left side of the brain, the location of the processes that govern reading, speaking, and calculating, is a tool of precision and analysis. It is the logical and rational part of the brain. In a complementary way, the right side of the brain is a tool of integration and synthesis. It enables one to recognize a pattern or a melody, and it controls the sequence of coordinated movements that one employs in sports or in dancing. It confers the sense of timing and it dominates artistic creation. Through the use of symbols, analogies, visual representations, and models, it is the framework of intuition.

We still cannot explain why evolution produced such a differentiation in the brain. But we recognize that our education seems to favor the left side of the brain disproportionately over the right side. That is, it favors analytical thought over systemic thought; rational thought is emphasized rather than intuitive thought. Doubtless at some time in the evolution of man and humanity the analytical, logical, and rational approach was one of the conditions for the survival of the species and for the domination of nature. This may no longer be true today.

The great constants of human nature are expressed as needs or drives at the biological, intellectual, social, and symbolic levels. But I prefer to speak here of *components* (which introduce the idea of properties) rather than traditional needs, which seem to be too closely linked to a given socioeconomic context. The four fundamental components are:

the biological, in which the organism is the unit; the intellectual and behavioral, in which the person is the unit; the social and relational, in which the citizen is the unit; and the symbolical, in which the being is the unit. These four components are integrated in the totality that is the multidimensional man. Systemic education must also take into account this multiplicity of human dimensions.

The Principles of Systemic Education

On the practical level, how can we formulate and then apply the basic rules of systemic education?

One experiment has served as my model in formulating such principles. This experiment, the Unified Science Study Program (USSP), was conducted at MIT between 1967 and 1972 and was then taken up by numerous American universities. I was part of the teaching staff. Its "guinea pigs" were a hundred volunteers, eighteen-year-old freshman students. The original particulars of the program were that the students would study fundamental subjects (mathematics, physics, chemistry, biology, the humanities) related to a multidisciplinary project chosen from a list prepared by the team of fifteen teachers. In his own way the student would arrange to do his bibliographical research and then his laboratory experiments. Courses were prepared in cooperation with the teaching staff; some students would teach other students, and there would be no formal examinations. The student could demonstrate his command of the subject matter in four ways: the preparation of a minithesis; an oral or written examination; a presentation before the staff and students; a proposal for study that included a justification of the pertinent materials and the subsidy required for continuing the research.

The program was divided into five interdependent levels: atomic, molecular, biological, social, and ecological systems. The dynamics of these systems would be taught through simulation and teaching games, facts through self-instruction techniques (quizzes and self-teaching guides). There would be a "dry" laboratory for physics studies and a "wet" laboratory for chemistry and biology. Finally, one day a week would be given to reevaluating the program in the presence of both staff and students.

The general approach of the new education is clearly that of the systemic approach discussed in the second chapter. I simply want to add a few considerations of a practical nature that apply particularly to the first chapter, which was of a more pedagogical nature.

1. Avoid the linear or sequential approach. The traditional approach consisted of treating A in detail in order to understand B, which was then studied in detail in order to approach C. One never knows what the teacher wants to achieve; one only hopes it will ultimately be useful.

On the contrary, the systemic approach in education involves returning several times *at different levels* to whatever is to be understood and assimilated. It approaches the subject matter through progressive steps. Following a spiral passage, student and teacher take a preliminary look at the entire subject in order to define it, to evaluate the difficulties and the unknown areas, and then return to it in greater detail, even at the risk of some repetition.

2. Beware of overly precise definitions that may polarize and dry up the imagination. A new concept or law ought to be studied from various angles and seen in many contexts. This leads to the mutual enrichment of concepts through indirect illumination rather than the automatic use of a definition.

3. Emphasize the importance of mutual causality, interdependence, and the dynamics of complex systems by stressing disciplines that integrate time and irreversibility, such as biology, ecology, and economics. Even at the elementary level, the bases of systemic education could be represented by descriptive models or reasoning models used in these disciplines; they would complement the traditional instruction of mathematics, physics, and chemistry.

4. Use themes of vertical integration, general themes that make it possible to integrate several disciplines and several levels of complexity around a central axis. This is what I have tried to do in the chapters on energy, information, and time. Depending on the relevant levels of knowledge, one can even use more specific themes.

Here are several examples taken from the natural sciences. Around the concept of *continental drift* it is possible to teach the complementary aspects of geography, geology, biology, and ecology; at a higher level, geophysics, paleontology, genetics, and climatology. Using *blood and hemoglobin* as a central theme, one can bring out many of the fundamental laws and principles of physics, organic chemistry, biochemistry, molecular biology, physiology, cybernetics, and genetics. The theme of *the origin of life* can bring together astrophysics, physical chemistry, geology, molecular biology, biochemistry, and the theories of evolution and ecology. With the help of a theme such as *farm products* (food or animals), one can integrate elements of microbiology, nourishment and diet, hygiene, the process of fermentation, and the prevention of illness.

5. Keep in mind that the acquisition of facts cannot be separated

from the understanding of the relationships that exist among them. This principle is valid for all levels of instruction; only the resources change and adapt themselves to the levels of knowledge.

The Methods of Systemic Education

The first rule of systemic education is to let the student learn *at his own speed* (the principle of self-learning). The resources used will vary with the level of instruction. One might employ self-teaching modules that are made up of questions and answers but differ from programmed textbooks in that they permit greater flexibility. One might even use a computer-assisted instruction program specialized in a specific application.

The learning of the course material is supplemented by teaching kits that make it possible to perform simple experiments. The kits may include slides, 8mm film loops synchronized with texts recorded on cassette tapes, games, and models. This is the multimedia "package." The audiovisual thus has its place in these packages, where it is part of a whole educational system.

The second rule is *interaction*; its most often used resource is simulation. Simulation is building a model of reality and making it function as though it represented one aspect of that reality (see p. 82). In education simulation can take several forms.

*Noninteractive simulation** is represented by films, especially animated films, that communicate the dynamic elements of a complex process (chemical reactions, physical and mechanical laws, the functioning of a machine, biological and social processes, industrial growth, etc.). A computer equipped with a graphic output or visual displays and a camera can produce animated films. These films result from the synthesis of images rather than the filming of a subject in reality. Computer-made animated films will no doubt have considerable educational impact in the coming years.

Simulated games, with or without the help of a computer, are a basic method of systemic education. Very generally, a game can be defined as an activity taking place between two or more decision makers who try to attain their objectives (win the game) while taking into account various constraints and limitations (the rules of the game). Thus the game is a model of processes and rules that correspond to real events, situations, and objectives.

The educational simulation game takes various forms ranging from games in cardboard boxes to business games that utilize computers. In

* The creation of educational films by students can be considered a form of interactive simulation.

the latter each player has a role that corresponds to a real-life function: general manager, financial or marketing director, foreman, and so on. He formulates strategies, trades with others, forms partnerships, makes decisions, and evaluates in real time the consequences of his action—thanks to the information that is fed back to him from his environment.

While classical instruction concentrates on the events themselves, simulation affords an ideal way to facilitate the perception of the dynamic relationships that exist among the elements of a complex system. It is very difficult—if not impossible—to describe in words, spoken or written, simultaneous and interdependent interactions. It is much easier to understand the rules of a game, whether it be football or bridge, when one has seen it played or tried to play it oneself.

Simulation games can train one to find intuitive solutions to complex problems, to perceive opposition, conflict, balance of power, delay. The design and construction of a game also offers considerable educational value; one must first carry out a detailed system analysis, then execute the model. Consequently one must interrelate variables and then question their limits and the effects of their interrelationships.

Today simulation games are used on many occasions in business schools, industries, universities, and elementary schools. The "case study," often used in preparatory courses in business management, is another kind of game, but it cannot take advantage of the complexity of interactions in real time.

It is likely that the new generation of miniature computers or microprocessors will help to develop new interactive educational games in a decentralized and easy-to-use form that makes use of the ordinary television set.

Finally, *computer simulation,* when it is possible, is the indispensable complement to the acquisition of facts. (Its principal advantages were listed on p. 82.) Experience shows that students who have made a model of the system they are studying are led, as in a game, to ask "good questions"—the limits of variability in the parameters of a system, the precision of basic data to be introduced into the model. Now they look for these facts in specialized books, when formerly they did not feel motivated by the traditional course that consisted of learning a collection of unrelated rules and facts without first having a use for them. Model building and the writing of computer simulation programs are therefore particularly educational, especially when the programming language and the symbolism used are simple.

Education must first teach young people to create rather than to copy faithfully what has been created by others. They must also learn to understand the role of time's duration, which is part of any new work and gives it its unique character and its value. Traditional education has too often neglected this fundamental point—that there is no true original creation without the integration of time.

Obviously the arts lend themselves better to creative activity than do the sciences or technology, where it is difficult to create something entirely new. Traditional arts such as painting, music, and poetry are now part of the curriculum, but the accent is chiefly on the fidelity of the *copy* rather than on the process of artistic *creation.*

More modern forms of art such as photography, film, and the production of synchronized slide tapes can also offer students the means of creating original programs and performances. Dance, choreography, film direction, and all forms of artisan creation develop the sense of harmony of form and movement, the sense of *timing,* the precision and certainty of action. These activities balance the role of the right side of the brain vis-à-vis that of the left side.

Education must also provide the means of relating what has been learned to the immediate environment, to society, and to the world. The reintegration of newly acquired knowledge in its human, social, or economic context tends to reinforce the sense of responsibility and social utility. This re-creates the bond between pure fact and the milieu that makes it meaningful and (as Piaget would say) "operative."

This concern for the connection between theory and practice, integrated into the curriculum, is now to be found in several universities and an increasing number of countries as *alternate instruction.* In the course of studying for their degrees, students take paying jobs in business, then return to their universities to complete their formal studies.

Some cities are trying to interest students in municipal activities in order to enhance their sense of social responsibility. The sorting of household wastes, the recycling of materials, and the restoration of natural sites harmed by industry allows them to participate in a useful social activity and to relate their efforts to the ecological cycles.

Other experiments, especially in Canada and in France, have given high school students the opportunity to prepare, with the help of portable television equipment, documentaries and newscasts for local stations and cable television.

One of the surest ways to master a new subject is to teach it to others. Thus students may teach the basics of a course to younger or even older students—and in this way instruction can "snowball."

This form of mutual instruction can be replaced or supplemented by a rather peculiar type that involves a teaching machine. When a student writes even a simple program for the computer (a very dumb machine that has to have everything explained to it), the student learns to be simple and specific and at the same time to generalize the facts, rules, and restraints that he prepares and puts into the machine. Abstract concepts that are often difficult to assimilate—concepts such as variables, equations, derivatives, asymptotes—take on their full meaning when the student succeeds in "teaching" them to the computer by programming it.

The new approach, then, tries to reverse the unidirectional flow of ideas from the teacher to the class. Now it is up to the students to organize their knowledge from component materials provided by the teacher. At present, paradoxically, it is the teacher who performs the most creative part of the work—in preparing the materials for the course. The students find themselves in the uncomfortable position of having to reassemble the pieces as accurately as possible.

Possible Structures of Parallel Education

The structure of education today takes the form of a tree; we acquire the basics by following a common trunk, then we specialize in order to find a career. Backstepping is impossible or certainly very difficult.

The structure of systemic education would take the form of a pyramid; we would enter at the top through what is most general, most common, even most intuitive in elementary education, then we would define in broad terms the goals of our own education and move toward the base of the pyramid to obtain essential knowledge. Ultimately, everything we learn would be related to action (or the simulation of action).

Are today's centralized structures of instruction ready for such an upheaval? And what methods might be used alongside traditional methods?

Mutual instruction. Each individual is the center of a communication network and consequently a potential source of knowledge or know-how. Mutual instruction on the most varied subjects will be developed and applied only at the level of small communities managed by the participants themselves—cultural associations, social clubs, business groups, and senior citizen associations. In the very long term its effectiveness might be greater with computer selective matching and "horizontal" communication networks.

The university without walls. This is already a reality for many people receiving their initial training or a continuing education. The way was led by the Open University in the United Kingdom. The development of cable television and videocassettes will probably enlarge on an educational process that is already bearing fruit and help to make it more widespread in many countries. Media other than radio and television are beginning to cooperate in long-distance instruction. In some California towns the local newspapers publish entire sections devoted to courses for which students can obtain credits at the nearest university or take their examinations.

Free access to knowledge. To extend the notion of self-service to all educational material it is not necessary to wait for the installation (perhaps only hypothetical) of computer data banks for public use or selective information access systems (described in the fourth chapter). Ordinary libraries clearly act as self-service centers for education. And some university libraries, especially in the United States, have an audiovisual center where one can find slides, film loops, tapes, models, and games, depending on the specific programs and courses. Such a center is transformed into a learning center when the student can project the films and slides he chooses. Sometimes students can interact with a computer controlling a multimedia terminal.

Why would it not be possible to have self-service stores devoted exclusively to educational products? It would be like filling a basket in a supermarket, but instead of food or soap powder the basket would contain educational "packages" of books, magazines, games, models, and audiovisual materials.

The computer center. Another form of free access, the computer center of many universities and engineering schools in the United States and (more recently) Europe, tends to become an "open house." The computers of the data processing department are available to students at night and on weekends as well as during the usual hours of courses. Students learn "on the spot," working with their own problems, as much about data processing as about their own fields. Thus they can quickly apply their knowledge and verify its range. The role of the more experienced student to whom the novice students address questions is also important. This education is empirical: the theory arrives afterward, the computer acting as a catalyst that accelerates the acquisition—and especially the integration—of knowledge in a greater, more comprehensive system.

Instantaneous feedback. Interaction in real time allows one to learn through trial and error with the help of feedback. The need for this led to the study and the installation of systems of interrogation, communication, and participation adapted to different environments. For some time there have been classrooms equipped with automatic response systems. These systems function by means of keyboards placed at students' desks; each of the several keys corresponds to one answer to a multiple-choice question asked by the teacher. All answers (and in some equipment the response times) are recorded on a terminal at the teacher's desk. Such techniques develop slowly, for they limit the students' choice and they depend too much on the way in which questions are phrased.

Commercial firms are already selling interactive participation systems to universities (for seminar courses, round-tables), businesses (for boards of directors, management meetings, workgroups), and international symposiums (for questions, comments, seminars). These mininetworks use miniaturized response terminals that can be held in the hand of each participant; a microprocessor and a giant screen make it possible to tabulate and display the results. In this way we have the use of a continuous opinion poll, made necessary by the increasing effectiveness of group work or collective creation.

Educational parks. These parks could be created by town councils in cooperation with private businesses. Planned in the manner of amusement parks or wildlife parks where animals wander in freedom, these parks would be an extension of the science museums. Young people could play while observing nature and participating in real experiments. Educational parks would try to avoid the diciplinary approach; they would bring together the exact sciences, the human sciences, and technology in order to emphasize not only their complementarity but the importance of a common ground where new methods and techniques could be used to try to resolve complex problems.

The role of industry. The privileged center of professional training and continuing education, industry is probably the level at which systemic education will come to occupy—and more quickly than elsewhere—a choice location. The global model that business represents lends itself particularly well to the description and the assimilation of basic systemic facts (see p. 33). It is not a question of adding a new discipline (such as data processing or marketing) but of learning to collect, integrate, and rank the quantities of information that come continuously from the environment in which one lives. It is thus a matter of synthesizing rather than absorbing ready-made "recipes," often poorly applied for lack of a general frame of reference.

Numerous educational seminars already solicit business personnel. Perhaps we should try to take advantage of the possibilities offered by the extension of flexible working hours. A special card (like that inserted in electronic terminals for recording work hours) would make it possible for members of a given company to earn an "education credit" by attending at times of their choice (lunch hour or off-peak hours) continuing courses taught right in the company. This education "canteen" is possible today through models of audiovisual classes that permit students to take these courses on a continuous basis, answering the programmed questions and thereby accumulating credits that can be recorded from week to week.

Senior citizen universities. In a society of growth and consumption there is little room for the aged. Youth seems to take precedence over the experience of age. The hiding of death in our society leads to the hiding of old age. But everything can be modified in a society of stationary population and economy. The decline of the birth rate in the developed countries is already changing the pyramid of age and leading to an increase in the population of the aged. Because of the appearance of new values introduced into a society with a stationary economy (such as respect for experience and for those things that endure and perpetuate themselves), we can expect that senior citizens will strengthen their position in tomorrow's world. Perhaps they will find again the happiness and respect that aged people enjoyed in ancient societies or the prestige of the elders in so-called primitive cultures.

In place of the open conflict between the generations that is customary today (accompanied by the scorn and selfishness of youth and the barriers raised by older people trying to defend their privilege) we must perhaps anticipate and prepare for an unprecedented cooperation between youth and the aged. As older people take advantage of the educational opportunities open to them, as they reflect on and synthesize newly learned material, their minds will accept the general ideas and global approaches with greater facility. This is the role that the universities for senior citizens must take up in order to arrive at a more equitable sharing of powers in a more balanced society.

Seven

Scenario for a World

I should like this final chapter of the book to be an opening onto the future, not a conclusion. Every criticism, every thorough examination of one type of society and its scale of values ought to lead us toward a new design for society. How can we discern the major features of this society through the gropings of social innovation—the experiments, the successes, the failures that we witness? From what point of view can we formulate and represent such a design?

I propose to reassemble in condensed form the principal themes of the preceding chapters. There are several ways of doing this. One can apply classical methods of forecasting and then try to describe in detail one aspect of the future society. One might, for example, project a small number of tendencies from among the most marked. Or one might adopt the "prospective" attitude, studying the present from the viewpoint of a desirable future in order to determine the meaningful events of today.

One can also try to confront the principal themes of the main currents of contemporary thought that I have presented by adopting a descriptive attitude, the most objective possible. Or, on the other hand, one might choose a normative attitude and orient the proposition in terms of a personal position or an ideology.

Beyond the normative and the objective there are also the expedients of science fiction, political fiction, and utopian writing. All these methods are well known to planners and futurologists and are widely used.

In terms of my own objective, however, the one method that appears to combine them advantageously is the method of "scenarios." The principle of that method is that the future is never given in its totality; it can be determined only through choices made by people devoted to building their future. Thus there is an infinite number of possible "futures," and a scenario is nothing but a more or less detailed description of some of them. A scenario clarifies decisions and facilitates choices.

But a scenario does not describe what is probable or even what is possible. For between the probable and the possible there is political will as much as there is chance, catastrophe, global crisis, or revolution. A scenario describes situations as they might be, situations that are plausible in a given context and in terms of what one knows of the evolutionary tendencies of the principal elements of the system under study. In this respect the scenario is quite like a game; one acts as though the description were possible and one had some relation to it.

Every scenario is a bit biased, as is the case with the present one. First because it is unique, whereas usually the rule insists that one compare several scenarios (for example, the pursuit of unrestrained growth; the slowing of economic growth while the present pursuit continues; catastrophes; the global crises of the economies; wars and other conflicts)—but such a comparison would take too long. Secondly because one again encounters several of the ideas, suggestions, and theses that I propose and defend in this book. (It will be easy for you to pick them out, recognize them, and criticize them.) My purpose, I recall, is to stimulate thought and reflection, not to attempt to impose my opinions. In order that you may use your imagination as you will, this scenario voluntarily assumes the somewhat dry form of an outline: I have conceived it in the form of notes sent by a reporter to a large weekly newsmagazine. The details are left to you to invent.

When will the scenario take place? Does it refer to a particular country or to a composite of several countries? It is neither possible nor even necessary to be precise. Some of the situations described in the scenario could exist in the 1980s, others not before the end of the century—and only in the so-called developed countries.

Travel Notes in Ecosocialism (August 12 to October 15, 8 A.C.*)

Ecosocialism, ecosociety, ecocitizen, ecocommunications, ecohealth, ecocongress. . . . This is not a new "ecocult"! The prefix "eco" symbolizes here the close relationship between economy and ecology; it puts the accent on *relationships* among men and between men and what they call their "home," the ecosphere.

At the time of the first electronic referendum taken on individual terminals the ecocitizens preferred, instead of a national anthem, a quotation from Dennis Meadows, an American university professor who in 1971 had called attention to the need for limiting growth.

> After two centuries of growth, we are now burdened, in the natural and social sciences, with blind decisions and obligations. At the present time, there is no economic theory of a society founded on technology where the rules of interest lead to zero, where the productive capital does not lead to accumulation, and where the main concern is about equality rather than growth. There is no sociology of balance which is interested in the social problems of a stabilized society, where men and women of an older age are in the majority. There is no political science of equilibrium capable of enlightening us on the means of exercising the democratic choice in a society where short-term material gain would cease to be the criterion of political success. There is no technology of balance which gives absolute priority to the recycling of all forms of material; to the use of solar energy which is not a pollutant; to the minimization of flows of material as well as energy. There is no psychology of the state of stability which lets man find a new image of himself or allows him to find other means of motivation in a system where material production would be constant and balanced according to the limited resources of the earth.
>
> This would be the great challenge to each of our traditional disciplines: to elaborate on the project of a society which finds its material motives and its attractiveness in a state of equilibrium. The task would be heavy with technical difficulties and concepts. The solutions would not only be more satisfactory to the spirit but would also be an immense advantage to society in general.

The coming of ecosociety took place in three main stages, each founded on a type of economy that corresponds to a given environment: the economy of survival (primitive society), the economy of growth (industrial society), and the economy of equilibrium (postindustrial society or ecosociety).

The economy of equilibrium (or stationary economy) that characterizes ecosociety today does not imply—as some believed in the late 1970s—

* A.C., after the crisis, or following the great worldwide crisis of the economies.

a "zero growth." The limiting of choice to two alternatives, "growth at any price" and "halting growth," was probably the result of the preponderant use of a logic of exclusion peculiar to that period, a type of logic that eliminated any nuance of meaning, any complementarity. It was obvious that the real question was not one of growing or not growing but rather the problem of how to *reorient* the economy to serve better *at the same time* human needs, the maintenance and evolution of the social system, and the pursuit of true cooperation with nature.

The economy of equilibrium that characterizes ecosociety is thus a "controlled" economy in the cybernetic sense of the term. Some sectors can pass through phases of growth, others are kept in dynamic equilibrium, and still others maintain a "negative" rate of growth. The "equilibrium" of the economy results from the harmony of the whole. As in life, this stationary state is a *controlled disequilibrium.*

One model of society proposed during the 1970s came close to ecosociety; this was the convivial society of Ivan Illich. But this model was also far from it when one considers certain aspects that I shall describe. First we must recall the meanings according to Illich of the two fundamental concepts of *conviviality* and *radical monopoly.*

> A society in which modern technologies serve politically interrelated individuals rather than managers, I will call "convivial." . . . I have chosen "convivial" as a technical term to designate a modern society of responsibly limited tools.

> The man who finds his pleasure and sense of balance in the use of the convivial operation is *austere.* Austerity does not have the connotation of isolating nor of enclosing oneself. Austerity according to Aristotle and to Saint Thomas Aquinas was founded on friendship.

> The establishment of radical monopoly happens when people give up their native ability to do what they can do for themselves and for each other, in exchange for something "better" that can be done for them only by a major tool. . . . This domination assures obligatory consumption and subsequently restrains the autonomy of each individual. It is a particular type of social control reinforced by the obligatory consumption of mass production that only the heavy industries can provide.

Illich in his model appears to have underestimated certain technologies whose development was slowed neither by crises nor by changes of government: the telecommunications explosion, the miniaturization and decentralization of data processing, and mankind's mastery of certain natural processes, particularly in biology and ecology. Telecommunications and microcomputers have thus permitted the creation of decentralized networks of "distributed knowledge" controlled by the users themselves. This progress had been made possible by a closer association between the human brain and the computer. This association, founded on voice recognition, handwriting recognition, pattern recognition, and a verbal dialogue with the computer, has gradually changed the computer into a veritable intellectual assistant.

The mastery and the imitation of some natural processes were achieved at the industrial level through the use of microorganisms and enzymes in the production of food, medicines, and chemical substances useful to society, and at the ecological level through the control and regulation of natural cycles with the objective of increasing agricultural production or eliminating more efficiently the wastes of social metabolism. These techniques of bioengineering and ecoengineering opened the way to new industrial processes that are less polluting, that use less energy, and that are easier to control and decentralize than were the old procedures of mass production.

Lenin used to say, "Communism is the Soviet people plus electricity." By the same token, *ecosociety is conviviality plus telecommunications!* For the great economic crises and the technological breakthroughs transformed the classical industrial society by means of a double movement: a decentralization (or differentiation) leading to the mastery and control of modern tools and a refocusing (or integration) resulting principally from progress in telecommunications and microcomputers.

This double movement fostered an increase in the effectiveness of community management at the base level (and consequently the progressive disappearance of certain "radical monopolies") and an increase in each individual's participation at all levels of the social system.

Decentralization is based on individual responsibilities, while participation allows a regulation of the metabolism of society (from the decentralized level to that of the great macroscopic feedback control loops). Clearly this reestablishment of the balance of powers is accompanied by deep modifications in the political, economic, and social structures.

Contrary to the industrial societies of the classical type, structured "from top to bottom," ecosociety is structured from "bottom to top," from the individual and his sphere of responsibilities through the organization of communities of consumers who guarantee the decentralized management of the principal organs of the life of the society—notably the energy transformation systems, the educational systems, and the electronic systems for communication, participation, information processing, and (in certain sectors of industry) production.

Ecosociety acknowledges the coexistence of private ownership and state ownership of production systems. In the extension of the liberal regime ecosociety favors innovation and the ability of free enterprise and free competition to adapt. However, it submits businesses to strict control by the communities of consumers and users. These communities work closely with political leaders at the national level through a participatory planning system that allows the selection of the major objectives and the determination of the principal deadlines.

"Social feedback," which takes place at all hierarchical levels of society, allows the control and the application of participatory planning as well as the adaptation to new conditions of evolution.

The main feedback controls apply for the most part to energy consumption, the investment rate, the population growth rate, and the principal cycles corresponding to the functions of supply, production, consumption, and recycling.

Energy consumption is maintained at the level that existed at the beginning of the 1980s. This is not monastic austerity; the energy is better distributed, better conserved, and more efficiently used.

Investments in new production capacity serve to balance the obsolescence of machines and buildings and to open up new areas according to social needs.

The birth rate is maintained at a level that equals the death rate of the population; this guarantees a stationary state.

The cycles of supply, production, consumption, and recycling were completely reorganized. The creation of channels of recovery and decentralized systems for sorting materials have enabled the metabolic cycles of the social organism to be connected again with the natural cycles of the ecosystem.

Ecosociety is decentralized, community-minded, participative; individual responsibility and initiative really exist. Ecosociety rests on the pluralism of ideas, styles, and ways of life. As a result equality and social justice are making progress, and there are changes in customs, ways of thinking, and morality. People have invented a different life-style in a society in equilibrium. They have realized that the maintenance of a state of equilibrium is more delicate than the maintenance of a state of continued growth.

With the help of a new vision, a new logic of complementarity, and new values, the people of ecosociety invented an economic doctrine, a political science, a sociology, a technology, and a psychology of the state of controlled equilibrium.

This other way of life is expressed in all social activities, especially in the organization of cities, work, human relationships, culture, customs, and manners. (The total integration of telecommunications in everyday life is significant here.)

The cities of ecosociety have been thoroughly reorganized. The oldest sections were restored to the people, free of automobiles. There the air is again fit to breathe and silence is respected. Pedestrian ways are numerous; on the streets and in the parks the people take their time.

The new cities are broken into multiple communities made up of interconnected villages. It is a "rural" society, one that is integrated through an extraordinary communications network that does away with needless travel and enables many people to work at home.

In business and industry many employees are no longer required to spend long hours at rigorous work. The extension of methods for managing working time has brought about a veritable liberation of time. The breaking up of individual hours and the synchronization of activities that results from it were balanced by the accountability of a "collective time" that permits a better distribution of work both in industry and in society. The management of time also affects other periods of life: vacation time, education, professional training, careers, and retirement.

Ecosociety catalyzes the appearance of service activities—the almost total dematerialization of the economy. A large percentage of social activities is based on mutual services and the exchange of services. The matching of people and ideas is facilitated by the new communications networks—intellectual endeavor through decentralized computer systems.

Industrial societies formerly were unable to support the exorbitant increase in the costs of education and health, and the quality of these services deteriorated. Ecosociety started again from the nodes of the human network. Mutual instruction and mutual medical assistance were achieved on a grand scale. Whereas the mastery of the megamachine of the industrial societies required an advanced education, specialized instruction in ecosociety is considerably reduced. It is now more global, more practical, and more meaningful. Meanwhile, people consume less drugs, call their doctors less often, and go to hospitals only in exceptional cases. Living is healthier, the methods of preventing illnesses more effective. More time is devoted to stimulating natural immunities than to controlling diseases by means of "outside" chemical agents. Balanced nutrition and exercise are key factors in self-management of health.

Oil and energy are still widely used in ecosociety, but their use has been stabilized at a level that permits an equitable distribution of resources. This has led to deep modifications. Programs for putting into operation new nuclear power centers have been dropped. The decentralization of energy transformation plants has led to the exploitation of new energy sources. Above all, energy conservation and the general struggle against waste have made it possible to stabilize energy consumption. Society has learned to use the internal energy of social systems, energy that was formerly expended only in periods of crisis—war or revolution.

Motivation that leads to action used to be inspired by self-interest (money, honors), by constraint (regimentation, fear of fines), and occasionally by the comprehension of the usefulness of one's action and a sense of social responsibility. The "transparency" of ecosociety, better information, and more effective participation have led gradually to the bringing into play of the two latter motivations, without which there is no real social cohesion.

In industry and farming the energy-intensive procedures were replaced by soft technologies and natural processes. In some transformation industries, such as petrochemistry, activities that had high energy costs were abandoned. The recycling of calories and raw materials is practiced on a wide scale. Manufactured products are more durable and easier to repair; thus maintenance and repair have become revitalized activities. Craftsmanship has been reborn, and objects are personalized rather than standardized.

The biotechnological revolution radically modified agriculture and the food processing industry. New bacterial species have become man's allies in production and recycling. Artificial enzymes are used to produce fertilizers and foods. But there are still restrictions because of the thoughtless waste of the previous industrial society.

Ecosociety is an explosion of quality and feeling, the exploration and conquest of inner space. Less preoccupied with economic growth, and producing and consuming less, people have again found time for themselves and for others. Human relationships are richer and less competitive; people respect the choices and freedoms of others. Everyone is free to pursue pleasure in all its forms: sexual, aesthetic, intellectual, athletic. Individual creativity and personal accomplishment play an important role in the community. People admire the unique and irreplaceable character of a work of art, a scientific discovery, or an athletic achievement.

Scientific progress was marked by the prodigious development of biology. Yet more than ever there are problems in the relationships between science and politics, science and religion, science and ethics. A "bioethic" reinforces the new morality of ecosociety. It is founded on respect for the human person; it orients and guides one's choices. For the people of ecosociety have amazing power at their disposal: hormonal and electronic manipulations of the brain, genetic manipulations, syntheses of the genes, chemical actions on the embryo, *in vitro* culture of the embryo, choice of sex, and control of the processes of aging.

The relationship between man and death has evolved; death is accepted and reintegrated into life. The aged participate in social activities, and they are the object of respect and consideration.

A religious feeling (an emergent religion, not merely a revealed religion) enriches all activities of ecosociety. It supports and validates action; it offers the hope that something can be saved because there exists in every one of us a unique power of creation and because the outcome of society rests in collective creation.

This is one scenario from among many, for one world among many. Is it a dream for the most part? Perhaps. But it is important to dream. And why cannot dreams be taken for realities . . . long enough to build a new world?

Paris, September 1978

Notes

Works that served as basic documentation or that would allow the reader to pursue a subject further are grouped immediately following the section title. The author (and date) listings refer the reader to the Bibliography, where the references are given in full.

All the diagrams are original except for those on pages 32(top), 101, 109, 189, which were adapted from Wolman (1965); *National Geographic Magazine,* November 1972; *Energy;* Time-Life Collection, "The World of Science"; Lehninger, 1969.

INTRODUCTION: THE MACROSCOPE

The term "megaloscope" was used by Lewis Carroll.

Macroscope is the title of a science fiction novel by Piers Anthony, published in 1969. Howard T. Odum (1971) also used the term "macroscope" in ecology.

ONE. THROUGH THE MACROSCOPE

Aguesse (1971), Clapham (1973), Ramade (1974).

Albertini (1971), Attali and Guillaume (1974), Perroux (1973).

13 The quotation from L. Robbins appears in Attali and Guillaume (1974), p. 9.

13 Passet (1974), p. 232.

14 Attali and Guillaume (1974), p. 10.

Sjoberg (1965), Laborit (1972), Forrester (1969).

31 The figures on metabolism in cities come from Wolman (1965) and Lowry (1967).

33 The term "megamachine" was used by Lewis Mumford (1974). The first definition is from Albertini (1971), p. 37; the second from Attali and Guillaume (1974), p. 27.

Nourse (1965), Laborit (1963) and (1968).

38 Schlanger (1971).

43 Walter Cannon (1929) and (1939). The comparison of plasma with the primitive ocean is from Laborit (1963).

Laborit (1963), J. de Rosnay (1965), Lehninger (1969), Watson (1972).

53 The symbolic representations of hemoglobin are from Perutz (1971). See also J. de Rosnay, "The function of hemoglobin," *La Recherche, 14,* 677, 1971.

TWO. THE SYSTEMIC REVOLUTION: A NEW CULTURE

General Systems Yearbook, beginning

1954; Young (1956), Ashby (1956), Ackoff (1960), Churchman (1968), Berrien (1968), von Bertalanffy (1968), Buckley (1968), Emery (1969), Barel (1971) and (1973).

58 The definition of the word "system," which occurs again on p. 65, is from Hall and Fagen (1956).

58 Wiener (1948), von Bertalanffy (1954) and (1968).

62 Shannon and Weaver (1949).

63–64 The first references to industrial dynamics are in Forrester (1958) and (1961).

64 Couffignal (1963), p. 23. The references to Plato and Ampère are in Guillaumaud (1965); see also Guillaumaud (1971).

69 The symbols of the structural and functional elements of a system are derived from those used by Forrester (1961). See also Meadows (1972).

73 The essential role of the flow variables and state variables was stressed by Forrester (1961), pp. 67–69.

77 A study of "world models" was made by Cole (1974); see also Mesarovic and Pestel (1974).

82–83 There is an excellent study of the advantages and the dangers of simulation in Popper (1973), pp. 40ff.

83 The allusion to "mental models" is in Meadows (1974).

87 On counterintuitive behavior of complex systems, see Forrester (1971).

87 The law of requisite variety was proposed by Ashby (1956); see also Ashby (1958).

90–91 The example of solid wastes is from Jorgan Randers (1973).

92 "To evolve, allow aggression." See also the role of "events" in the evolution of a complex system in Morin (1972).

95 An excellent critique of the systemic approach and its fecundity appears in Morin (1977).

95 Letter from Engels to Lavrov in Marx and Engels (1973), p. 83.

96 The term "noosphere" is from Teilhard de Chardin (1957).

THREE. ENERGY AND SURVIVAL

97 On the relationship between bioenergetics and ecoenergetics, see J. de Rosnay (1974).

Puiseux (1973) (who quotes from the works of A. Varagnac), Illich (1973), Leroi-Gourhan (1972).

104 The law of "maximum energy" was proposed by Lotka (1956).

105 The examples that illustrate the law of optimum yield are from Odum (1955).

105 The school of "thermodynamics of irreversible processes" includes Onsager, de Groot, de Donder, Prigogine (1969) and (1972). See also Katchalsky (1971) on network thermodynamics.

Matthews et al. (1971).

108–110 The statistics are from several publications, among them Cook (1971), Ramade (1974).

112 The observations of the *Bulletin of the World Meteorological Organization* are cited by Kukla (1974). The effects of atmospheric dusts are studied in detail in Hobbs et al. (1974) and Bryson (1974).

114 The table of values in kilocalories was compiled from several sources, among them Slesser (1973), Odum (1971), Hannon (1974).

114 On the energy equivalent of the kilocalorie, see Odum (1974), p. 46.

115 On energy analysis, see Slesser (1973), Berry (1974), Hannon (1974).

116 The estimate of energy costs in producing a car is from Berry (1974).

117 See J. de Rosnay, *La Recherche, 47,* 695, 1974.

117 On the expenditure of energy to feed the United States, see Hirst (1974).

118 Application of energy analysis to agriculture: Pimentel (1973), Steinhart (1974).

120 Competition between energy and work: Bezdek and Hannon (1974).
124 On the manufacture of fertilizers through using nitrogen, see J. de Rosnay, "Toward a bioindustry of ammonia," *La Recherche, 32,* 278, 1973.
126 On immobilized enzymes, see Zaborsky (1973).
129 The expression "postindustrial society" is used by Touraine (1969) and Bell (1973).

FOUR. INFORMATION AND THE INTERACTIVE SOCIETY

132 The "theory of information" was principally the result of the work of Hartley (1928), Szilard (1929), Gabor (1946), Shannon and Weaver (1949), Brillouin (1959).
132 The example of the card game was suggested by examples in Brillouin (1959) and Costa de Beauregard (1963), p. 63.
134 The example of the reading of a printed page was suggested by Tribus (1971).
137 The expressions "planetary village" and "global village" are from McLuhan (1965).
138 The expression "society in real time" was proposed by the author in J. de Rosnay (1972).
141 The expression "left out of power" was used by J. Attali in a report to the Group of Ten on social malaise. The report has not been published.
142 The data used in the preparation of the section on communications hardware were taken chiefly from Sprague (1969), Parker (1969) and (1972), Martin (1971), Dickson (1973).
144 The examples of services in real time came from studies made in the United States by the author and from Goldmark (1969), Dickson (1973), Martin (1971), National Academy of English Report (1971), Walker (1971), Day (1973).
147 On substituting communications for travel, see Goldmark (1971), Dickson (1971), Day (1973), National Academy of English Report (1971), Attali (1974).
149 Friedman (1974).
Leonard and Etzioni (1971), Stevens (1971) and (1972), de Sola Pool (1971) and (1974), Singer (1973), Carroll (1974).
149 The term "social feedback" is proposed to emphasize the cybernetic nature of information feedback loops. Several authors use the terms "citizen feedback" (Stevens, 1971), "instant democracy," and "participatory democracy." See also the excellent examples of participative democracy in Jungk (1974), pp. 157ff.
156 These instruments are sold commercially by Applied Futures Inc., Greenwich, Conn., or used at MIT by Prof. Thomas B. Sheridan.

FIVE. TIME AND EVOLUTION

Gold (1965), Blum (1962), Costa de Beauregard (1963b), Berger (1964).
163 Bergson (1948), Teilhard de Chardin (1957).
164 Costa de Beauregard (1963a), Grunbaum (1962) and (1963), Reichenbach (1956), Gal-Or (1972).
164 On the relationship between information and entropy, see also Atlan (1973), Laborit (1974).
166 The distinction between time that "spreads out" and time that "adds on" is from Saint-Exupéry.
170 Grunbaum (1962), Costa de Beauregard (1963).
The main elements of this chapter appeared in J. de Rosnay (1965).
172 Teilhard de Chardin (1955), Monod (1970).
174 The term "integron" is from Jacob (1970); "holon" is from Koestler.
177 Bergson (1948). The story of the "train of the second principle" appears in J. de Rosnay (1965).

On the general mechanisms of evolution, see Von Foerster (1960), Prigogine (1972), Eigen (1971), Monod (1970), Atlan (1972), Morin (1973).

182 The relationship between autocatalysis and biological reproduction was stressed by Calvin (1956).

185 The note refers to works in prebiotic chemistry; see J. de Rosnay (1966).

186 On acceleration, see also Meyer (1974).

187 Darwin (1959), p. 120; the correspondence between Marx and Engels is in Marx and Engels (1973).

188 Le Châtelier (1888).

189 The term "stationary economy" is from Daly (1973) and Boulding (1966).

189 Dupuy (1975).

189–190 The definition of a deadline with respect to the temporal dimension of the goal is from Idatte (1969).

190 On psychological time, see Lecomte du Noüy (1936).

SIX. VALUES AND EDUCATION

195 On acceleration, see Meyer (1974).

204 Reich (1970).

204 Garaudy (1971).

209 The quotations "a carbon copy of reality," "operative process," and "another verbalization of a picture" are from Piaget (1969), pp. 107, 110.

214 The definition of a game is from Abt (1970).

SEVEN. SCENARIO FOR A WORLD

223 Meadows (1974), pp. 63–64.

224 Illich (1973), pp. xiv–xv, 54.

Bibliography

Abt, Clark C., *Serious Games,* New York: Viking Press, 1970.

Ackoff, R. L., "Systems, organizations and interdisciplinary research," *General Systems Yearbook, 5,* 1960.

Aguesse, P., *Keys to Ecology,* Paris: Seghers, 1971.

Albertini, J. M., *The Wheels of the National Economy,* Paris: Workers' Edition, 1971.

Ashby, W. R., "General systems theory as a new discipline," *General Systems Yearbook, 3,* 1958.

———, "Requisite variety and its implications for the control of complex systems," *Cybernetica, 1,* 2, 83, Namur, 1958.

———, *Introduction to Cybernetics,* Paris: Dunod, 1958.

Atlan, H., *Biological Organization and the Theory of Information,* Paris: Hermann, 1972.

Attali, J., "A substitute for energy: communications," *Le Monde,* February 22, 1974.

———, *The Word and the Tool,* Paris: University Presses of France, 1975.

———, and Guillaume, M., *Antieconomics,* Paris: University Presses of France, 1974.

Barel, Y., "Prospective and systems analysis," *La Documentation française,* Paris, 1971.

———, *Social Reproduction,* Paris: Anthropos, 1973.

Bell, D., *The Coming of Post-industrial Society,* New York: Basic Books, 1973.

Berger, G., *Modern Man and His Education,* Paris: University Presses of France, 1962.

———, *Phenomenology of Time and Prospective,* Paris: University Presses of France, 1964.

Bergson, H., *Creative Evolution,* Paris: University Presses of France, 1948.

Berrien, F. Kenneth, *General and Social Systems,* New Brunswick, N.J.: Rutgers University Press, 1968.

Berry, R. S., and Makino, Hiro, "Energy thrift in packaging and marketing," *Technology Review, 76,* 32, 1974.

Bertalanffy, L. von, *General System Theory,* New York: Braziller, 1968.

Bezdek, R., and Hannon, B., "Energy, manpower, and the highway trust fund," *Science, 185,* 669,1974.

Boulding, K. E., "Toward a general theory of growth," *General Systems Yearbook, 1,* 1956.

———, "The economics of the spaceship earth," in Henry Jarret, ed., *Environmental Quality in a Growing Economy,* Baltimore: Johns Hopkins, 1966.
Brillouin, L., *Science and the Theory of Information,* Paris: Masson, 1959.
Bryson, R. A., "A perspective on climatic change," *Science,184,* 753, 1974.
Buckley, W., *Modern Systems Research for the Behavioral Scientist,* Chicago: Aldine, 1968.
Calvin, M., *American Scientist, 44,* 248, 1956.
Cannon, W. B., "Organization for physiological homeostasis," *Physiological Review, 9,* 1929.
———, *The Wisdom of the Body,* New York: Norton, 1939.
Carroll, J. M., and Tague, J., "The people's computer, an aid to participatory democracy," IFIPS conference proceedings, Human choice and computers, Vienna, 1974.
Le Châtelier, H., "Experimental and theoretical research on chemical equilibriums," *Annals of Mines,* Paris: Dumond, 1888.
Churchman, C. W., *The Systems Approach,* New York: Dell, 1968.
Clapham, W. B. Jr., *Natural Ecosystems,* New York: Macmillan, 1973.
Cole, S., "World models, their progress and applicability," *Futures, 201,* June 1974.
Cook, E., "The flow of energy in an industrial society," *Scientific American, 224,* 134, 1971.
Costa de Beauregard, O., *The Concept of Time,* Paris: Hermann, 1963.
———, *The Second Principle of the Science of Time,* Paris: Seuil, 1963.
Couffignal, L., *Cybernetics,* "What do I know?" series, Paris: University Presses of France, 1963.
Daly, H. E., *Toward a Steady State Economy,* San Francisco: Freeman, 1973.
Darwin, C., *The Autobiography of Charles Darwin,* New York: Harcourt Brace Jovanovich, 1959.
Darwin, F., *The Life and Letters of Charles Darwin,* New York: Basic Books, 1959.
Day, L. H., "An assessment of travel/communications substitutability," *Futures, 559,* December 1973.
Dickson, E. M., and Bowers, R., *The Video Telephone,* Ithaca, N.Y.: Cornell University, 1973.
Dupuy, J.-P., *Social Value and Congestion of Time,* Paris: CNRS, 1975.
Eigen, M., "Self-organization of matter and the evolution of biological macromolecules," *Die Naturwissenschaften, 58,* October 10, 1971.
Emery, F. E., *Systems Thinking,* London: Penguin, 1969.
Von Foerster, H., "On self-organizing systems and their environments," in Yovitz and Cameron, *Self-organizing Systems,* London: Pergamon Press, 1960.
Forrester, J. W., "Industrial dynamics: a major breakthrough for decision makers," *Harvard Business Review,* July–August 1958.
———, *Industrial Dynamics,* Cambridge, Mass.: MIT Press, 1961.
———, *Urban Dynamics,* Cambridge, Mass.: MIT Press, 1969.
———, *World Dynamics,* Cambridge, Mass.: MIT Press, 1970.
———, "Counterintuitive behavior of social systems," *Technology Review, 73,* 52, 1971.
Friedman, Y., *How to Live with Others without Being Chief or Slave,* Paris: Pauvert, 1974.
Gal-Or, B., "The crisis about the origin of irreversibility and time anisotropy," *Science, 176,* 11, 1972.

Garaudy, R., *The Alternative,* Paris: Laffont, 1971.

Georgescu-Roegen, N., *The Entropy Law and the Economic Process,* Cambridge, Mass.: Harvard University Press, 1972.

Gold, T., "The arrow of time," in *Time: Selected Lectures,* London: Pergamon, 1965.

Goldmark, P. C., "Telecommunications for enhanced metropolitan function and form," Report to the director of telecommunications management under contract no. OEP-SE-69-101, National Academy of Engineering, Washington, D.C., 1969.

———, "New applications of communications. Technology for realizing the new rural society," *World Future Society Conference,* 1971.

Grunbaum, A., "Temporally asymmetric principles," *Philosophy of Science, 29,* 1962.

———, *Philosophical Problems of Space and Time,* New York: Knopf, 1963.

Guillaumaud, J., *Cybernetics and Dialectical Materialism,* Paris: Social Editions, 1965.

———, *Norbert Wiener and Cybernetics,* Paris: Seghers, 1971.

Hall, A. D., and Fagen, R. E., "Definition of a system," *General Systems Yearbook, 1,* 18, 1956.

Hannon, B., "Options for energy conservation," *Technology Review, 76,* 24, 1974.

Hirst, E., "Food-related energy requirements," *Science, 184,* 134, 1974.

Hobbs, P. V., Harrison, H., and Robinson, E., "Atmospheric effects of pollutants," *Science, 183,* 1974.

Idatte, P., *Keys to Cybernetics,* Paris: Seghers, 1969.

Illich, I., *Deschooling Society,* New York: Harper & Row, 1971.

———, *Tools for Conviviality,* New York: Harper & Row, 1973.

Jacob, F., *Living Logic,* Paris: Gallimard, 1970.

Jungk, R., *A Wager on Man,* Paris: Laffont, 1974.

Katchalsky, A., "Network thermodynamics," *Nature, 234,* 393, 1971.

Kukla, G. J., and Kukla, H. J., "Increased surface albedo in the northern hemisphere," *Science, 183,* 709, 1974.

Laborit, H., *Biology and Structure,* "Ideas" series, Paris: Gallimard, 1968.

———, *Man and the City,* Paris: Flammarion, 1972.

———, *The New Grid.* Paris: Laffont, 1974.

Lecomte du Noüy, *Time and Life,* Paris: Gallimard, 1936.

Lehninger, A. L., *Bioenergetics,* Paris: Ediscience, 1969.

Leonard, E., Etzioni, A., et al., "Minerva: a participatory system," *Bulletin of the Atomic Scientists,* November 1971.

Leontieff, V., "The structure of the U.S. economy," *Scientific American, 212,* 25, 1965.

Leroi-Gourhan, A., *Actions and Words,* Paris: Albin Michel, 1972.

Lotka, A. J., *Elements of Physical Biology,* New York: Dover, 1956.

Lowry, W. P., "The climate of cities," *Scientific American, 217,* 15, 1967.

McLuhan, M., *Understanding Media: The Extensions of Man,* New York: McGraw-Hill, 1965.

Marcuse, H., *The One-dimensional Man,* Paris: Minuit, 1968.

Martin, J., *Future Developments in Telecommunications,* Englewood Cliffs, N.J.: Prentice-Hall, 1971.

Marx, K., and Engels, F., *Letters on the Natural Sciences,* Paris: Social Editions, 1973.

Matthews, W. H., et al., *Man's Impact on the Global Environment,* Cambridge, Mass.: MIT Press, 1971.
Mead, M., *Culture and Commitment: a Study of the Generation Gap,* New York: Doubleday, 1970.
Meadows, D. L., et al., *The Limits to Growth,* Potomac Associates, New York: Universe Books, 1972.
———, *What Limits?* Paris: Seuil, 1974.
Mesarovic, M., and Pestel, E., *Strategy for Tomorrow,* Paris: Seuil, 1974.
Meyer, F., *An Overheating of Growth,* Paris: Fayard, 1974.
Monod, J., *Chance and Necessity,* New York: Knopf, 1971.
Morin, E., "The event," *Communications, 18,* Paris: Seuil, 1972.
———, *The Lost Paradigm: Human Nature,* Paris: Seuil, 1973.
———, *The Method,* I: *The Nature of Nature,* Paris: Seuil, 1977.
Mumford, L., *The Myth of the Machine,* Paris: Fayard, 1974.
National Academy of Engineering, Washington, D.C., "Communication technology for urban improvement," June 1971.
Nourse, A. E., *The Body,* "The World of Science" series, New York: Time-Life, 1965.
Odum, H. T., "Trophic structure and productivity of Silver Springs, Florida," *Ecological Monograms, 27,* 55, 1957.
———, *Environment, Power and Society,* New York: Wiley, 1971.
———, and Pinkerton, R. C., "Time's speed regulator: the optimum efficiency for maximum power output in physical and biological systems," *American Scientist, 43,* 331, 1955.
Parker, E. B., "Some political implications of information utilities," *Conference on Information Utilities and Social Choice,* University of Chicago, December 2–3, 1969.
———, and Dunn, D. A., "Information technology: its social potential," *Science, 176,* 1392, 1972.
Passet, R., *Economics and the Living,* Paris: Melanges Garrigou-Lagrange, 1974.
Perroux, F., *Power and the Economy,* Paris: Bordas, 1973.
Perutz, M., "Hemoglobin: the molecular lung," *New Scientist and Science Journal, 676,* June 1971.
Piaget, J., *Psychology and Pedagogy,* Paris: Denoël, 1969.
Pimentel, D., "Food production and the energy crisis," *Science, 182,* 443, 1973.
Popper, J., *System Dynamics,* Paris: Organization Editions, 1973.
Prigogine, I., "Structure, dissipation and life," in Marois, ed., *Theoretical Physics and Biology,* Amsterdam: North Holland, 1969.
———, "The thermodynamics of life," *Research, 24,* 547, 1972.
Puiseux, L., *Energy and Confusion,* Paris: Hachette, 1973.
Ramade, F., *Elements of Applied Ecology,* Paris: Ediscience, 1974.
Randers, J., "The dynamics of solid waste generation," in D. Meadows, ed., *Toward Global Equilibrium,* Cambridge, Mass.: Wright-Allen, 1973.
Reich, Charles A., *The Greening of America,* New York: Bantam, 1970.
Reichenbach, H., *The Direction of Time.* University of California Press, 1956.
Revel, J.-F., *Neither Marx nor Jesus,* Paris: Laffont, 1970.
Rosnay, J. de, "Evolution and time," *Main Currents in Modern Thought, 27,* 2, November 1970.
———, *The Origins of Life,* Paris: Seuil 1966.
———, "Social systems in real time," *Proceedings of the Sixth International Congress on Cybernetics,* International Association of Cybernetics, 1972.

———, "From bioenergetics to ecoenergetics," *Communications, 22,* Paris: Seuil, 1974.
Schlanger, J. E., *Metaphors of the Organism,* Paris: Philosophical Library, 1971.
Shannon, C. E., and Weaver, W., *The Mathematical Theory of Communication,* Urbana: University of Illinois Press, 1949.
Singer, B. D., *Feedback in Society,* Lexington, Mass.: Lexington Books, 1973.
Sjoberg, G., "The origin and evolution of cities," *Scientific American, 213,* 54, 1965.
Slesser, M., "Energy analysis in policy making," *New Scientist, 328,* November 1973.
Sola-Pool, I. de, *Talking Back: Citizen Feedback and Cable Technology,* Cambridge, Mass.: MIT Press, 1974.
Steinhart, J. S., and Steinhart, C. E., "Energy use in the U.S. food system," *Science, 184,* 307, 1974.
Stevens, C. H., "Citizen feedback and societal systems," *Technology Review, 73,* 38, 1971.
———, "Towards a network of community information exchanges," in P. C. Ritterbush, ed., *Let the Entire Community Become Our University,* Washington, D.C.: Acropolis, 1972.
Teilhard de Chardin, P., *The Phenomenon of Man,* Paris: Seuil, 1955.
———, *The Vision of the Past,* Paris: Seuil, 1957.
Toffler, A., *Future Shock,* New York: Random House, 1970.
Touraine, A., *The Postindustrial Society,* Mediations Library, Paris: Denoël, 1969.
Tribus, M., and McIrvine, E. C., "Energy and information," *Scientific American, 224,* 179, 1971.
Vasse, Denis, *The Time of Desire,* Paris: Seuil, 1974.
Volterra, V., *Lessons on the Mathematical Theory of the Struggle for Existence,* Paris: Villars, 1931.
Watson, J., *Molecular Biology of the Gene,* Paris: Ediscience, 1972.
Wiener, N., *Cybernetics,* Paris: Hermann, 1948.
———, *Cybernetics and Society,* Paris: Union Generale d'Editions, 1962.
Wolman, A., "The metabolism of cities," *Scientific American, 213,* 178, 1965.
Zaborsky, O., *Immobilized Enzymes,* Cleveland: CRC Press, 1973.

Index